# Sustainable Energy Systems Planning, Integration and Management

Edited by

**Reetu Malhotra**
*Department of Applied Sciences*
*Chitkara University Institute of Engineering & Technology*
*Chitkara University*
*Rajpura 140401, Punjab, India*

**Pooja Mahajan**
*Department of Applied Sciences*
*Chitkara University Institute of Engineering and Technology*
*Chitkara University*
*Rajpura 140401, Punjab, India*

**Manish Singla**
*Department of Interdisciplinary Courses in Engineering*
*Chitkara University Institute of Engineering & Technology*
*Chitkara University*
*Rajpura 140401, Punjab, India*

&

**Ramesh Kumar**
*Department of Electronics and Communication Engineering*
*Manipal University Jaipur*
*Jaipur 303007, Rajasthan, India*

**Sustainable Energy Systems: Planning, Integration and Management**

Editors: Reetu Malhotra, Pooja Mahajan, Manish Kumar Singla and Ramesh Kumar

ISBN (Online): 979-8-89881-480-9

ISBN (Print): 979-8-89881-481-6

ISBN (Paperback): 979-8-89881-482-3



Published by Bentham Science Publishers Pte. Ltd. Singapore, in collaboration with Eureka Conferences, USA. 

First published in 2026.

need for a court order if at any point you breach any terms of this License Agreement. In no event will any delay or failure by Bentham Science Publishers in enforcing your compliance with this License Agreement constitute a waiver of any of its rights.

3. You acknowledge that you have read this License Agreement, and agree to be bound by its terms and conditions. To the extent that any other terms and conditions presented on any website of Bentham Science Publishers conflict with, or are inconsistent with, the terms and conditions set out in this License Agreement, you acknowledge that the terms and conditions set out in this License Agreement shall prevail.

**Bentham Science Publishers Pte. Ltd.**
No. 9 Raffles Place
Office No. 26-01
Singapore 048619
Singapore
Email: subscriptions@benthamscience.net

BENTHAM
SCIENCE

# CONTENTS

# FOREWORD

It is with great pleasure that I introduce the book, 'Sustainable Energy Systems: Planning, Integration, and Management.' This revolutionary work explores the transformative potential of sustainable energy in revolutionizing various sectors. Sustainable energy systems have emerged as a cornerstone in addressing global challenges, reshaping conventional paradigms, and unlocking new possibilities. From the inception of energy solutions to their evolution into advanced, integrated systems; this book provides a comprehensive exploration of the field's development.

The authors adeptly lay a strong foundation, ensuring accessibility for readers with diverse backgrounds as they delve into the fundamental principles and design methodologies underlying sustainable energy systems. Furthermore, the book explores advanced simulation techniques and tools that have become indispensable in modeling and forecasting energy system behavior. These cutting-edge methods enable researchers and engineers to unravel the complexities of sustainable energy systems, transcending the limitations of traditional experimental approaches.

This insightful narrative highlights the pivotal role of simulation and integration in optimizing sustainable energy system performance, unlocking their full potential for global impact. As you embark on this enlightening journey, absorb the wealth of knowledge and expertise encapsulated within these pages. Whether you are a seasoned practitioner, an aspiring innovator, or simply curious about the future of sustainable energy; this book promises to broaden your horizons and empower you to contribute to the advancement of this vital field.

**Ashok K. Chitkara**
Chancellor
Chitkara University
Punjab & Himachal Pradesh
India

# PREFACE

The field of engineering has always been driven by innovation and the ability to translate ideas into perceptible solutions. In recent years, one of the most transformative forces driving this innovation has been the advancement of energy. This book, Sustainable Energy Systems Planning, Integration and Management, delves into this exciting intersection, exploring how cutting-edge imaging techniques are revolutionizing diverse sectors. Sustainable Energy Systems in engineering have transcended their role in the energy sector. They now offer powerful tools for non-destructive testing, material characterization, process creation, and design optimization. This book provides a comprehensive overview of these advancements, highlighting their impact on various energy sectors. By fostering communication and knowledge exchange between specialists and engineers, we can unlock the full potential of these technologies and accelerate innovation across diverse fields. Throughout the book, we explore the fundamental principles of Sustainable Energy Systems, delve into their practical example in different sectors, and discuss emerging trends that promise to revolutionize the field further.

**Reetu Malhotra**
Department of Applied Sciences
Chitkara University Institute of Engineering & Technology
Chitkara University
Rajpura 140401, Punjab, India

**Pooja Mahajan**
Chitkara University Institute of Engineering & Technology
Chitkara University
Rajpura 140401, Punjab, India

**Manish Singla**
Department of Interdisciplinary Courses in Engineering
Chitkara University Institute of Engineering & Technology
Chitkara University
Rajpura 140401, Punjab, India

&

**Ramesh Kumar**
Department of Electronics and Communication Engineering
Manipal University Jaipur
Jaipur 303007, Rajasthan, India

# List of Contributors


| | |
|---|---|
| **Abhishek Kandwal** | School of Chips, Xi'an Jiaotong-Liverpool University, Taicang 215400, Suzhou, China |
| **Chander Prabha** | Department of Computer Science & Engineering, Chitkara University Institute of Engineering and Technology, Chitkara University, Rajpura 140401, Punjab, India |
| **Chetan Chauhan** | Department of Chemistry, Sardar Patel University, Mandi 175001, Himachal Pradesh, India |
| **Dipali K. Kolhe** | Department of Chemistry, Jaywantrao Sawant College of Engineering Hadapsar, Pune 411028, Maharashtra, India |
| **Gaganpreet Kaur** | Department of Computer Science & Engineering, Chitkara University Institute of Engineering and Technology, Chitkara University, Rajpura 140401, Punjab, India |
| **Ganesh Sakharam Chavan** | Faculty of Management Studies, Parul University, Vadodara 391760, Gujarat, India |
| **Himanshi** | Department of Computer Science & Engineering, Chitkara University Institute of Engineering and Technology, Chitkara University, Rajpura 140401, Punjab, India |
| **Jatin Arora** | Department of Computer Science & Engineering, Chitkara University Institute of Engineering and Technology, Chitkara University, Rajpura 140401, Punjab, India |
| **Kashmiri Khamkar** | Department of Chemistry, MIT World Peace University Kothrud, Pune 411038, Maharashtra, India |
| **Nidhi Deepak Sharma** | Department of Chemistry, Jaywantrao Sawant College of Engineering Hadapsar, Pune 411028, Maharashtra, India |
| **Nidhi Vijay Sharma** | Department of Chemistry, AISSMS's Institute of Information Technology, Pune 411001, Maharashtra, India |
| **Pawan Kumar** | School of Physics and Materials Science, Shoolini University, Solan 173229, Himachal Pradesh, India |
| **Pooja Mahajan** | Department of Applied Sciences, Chitkara University Institute of Engineering and Technology, Chitkara University, Rajpura 140401, Punjab, India |
| **Pradeep Ramesh Sonar** | School of Management, D Y Patil University, Pune 410506, Mahahrashtra, India |
| **Raman Kumar** | Department of Physics, Government College Daulatpur Chowk, Una 177204, Himachal Pradesh, India |
| **Reetu Malhotra** | Department of Applied Sciences, Chitkara Institute of Engineering & Technology, Chitkara University, Rajpura 140401, Punjab, India |
| **Sachin Trambak Mahale** | School of Management, D Y Patil University, Pune 410506, Mahahrashtra, India |
| **Sahil Garg** | Department of Mechanical Engineering, Chitkara University Institute of Engineering and Technology, Chitkara University, Rajpura 140401, Punjab, India |
| **Senthil Kumar AV** | Department of Computer Applications, Hindusthan College of Arts and Sciences, Udayampalayam 641028, Tamil Nadu, India |
| **Shilpy Bhullar** | Schepens Eye Research Institute, Massachusetts Eye and Ear Infirmary of Harvard Medical School, , Boston 02114, MA, USA |

**Sushil Kumar** School of Core Engineering, Shoolini University, Solan 173229, Himachal Pradesh, India

**Tanya Sood** School of Physics and Materials Science, Shoolini University, Solan 173229, Himachal Pradesh, India

**Ujjwal Kaushik** Chitkara College of Pharmacy, Chitkara University, Rajpura 140401, Punjab, India

**Yugen Sevda** Department of Computer Science and Engineering, Chitkara University Institute of Engineering and Technology, Chitkara University, Rajpura 140401, Punjab, India

# CHAPTER 1

# AI-Enabled Approaches for Optimizing the Smart Grid System

**Sahil Garg**[1], **Yugen Sevda**[2] and **Pooja Mahajan**[3,*]

[1] *Department of Mechanical Engineering, Chitkara University Institute of Engineering and Technology, Chitkara University, Rajpura 140401, Punjab, India*

[2] *Department of Computer Science and Engineering, Chitkara University Institute of Engineering and Technology, Chitkara University, Rajpura 140401, Punjab, India*

[3] *Department of Applied Sciences, Chitkara University Institute of Engineering and Technology, Chitkara University, Rajpura 140401, Punjab, India*

**Abstract:** A smart grid represents a huge leap forward from conventional grids and uses advanced technologies to make the power grid more efficient, reliable, and sustainable. One of its main features is two-way communications, which helps in self-healing, active consumer participation in managing energy, and resilience against both physical and cyber threats. A very important dimension of this transition is the integration of different sources of energy, including renewable energy, which plays a crucial role in reducing dependence on fossil fuels and maintaining an environmental-friendly energy path. The chapter examines key areas where AI can make significant contributions, including demand response management, energy forecasting, asset management, and fault detection. Predictions in the energy domain benefit from AI's processing power for huge volumes of data and continue to improve because of better accuracy, which will lead to grid stability. AI-powered fault detection systems can quickly identify and resolve issues, minimizing downtime and improving overall grid reliability. However, integrating AI into smart grids introduces new challenges, especially concerning cybersecurity and the increasing reliance on digital infrastructure. It offers the ability to optimize grid efficiency, integrate renewable energy more smoothly, and enhance resilience. This chapter contributes to the understanding of the potential benefits and challenges related to AI integration in smart grid systems, highlighting key approaches for future research and development in this rapidly evolving field.

**Keywords:** AI integration, Demand Side Management, Grid resilience, Renewable energy integration, Smart metering.

* **Corresponding author Pooja Mahajan**: Department of Applied Sciences, Chitkara University Institute of Engineering and Technology, Chitkara University, Rajpura 140401, Punjab, India; E-mail: pooja.mahajan@chitkara.edu.in

**Reetu Malhotra, Pooja Mahajan, Manish Singla & Ramesh Kumar (Eds.)**

# INTRODUCTION

With an increase in global population, the demand for electricity is rising. The world's electricity consumption has doubled over the past 30 years, reaching an estimated 28-thousand-Terawatt hours in 2021; however, the population is estimated to have increased by 8 billion [1] [2]. Electricity plays a crucial role in our everyday lives, especially in developed nations. The average household in the United States uses approximately 10,715 kilowatt-hours of electricity each year [3]. As developing countries rapidly industrialize, their energy needs are exploding.

Electric grids have traditionally served as the blueprints for power distribution systems in the past decades [4]. These grids are the coordinated structures engineered to deliver electric power from the generating facilities to electricity consumers. The backbone component consists of power generation facilities, high-voltage transmission lines, and distribution networks [5]. Electricity is produced at a power generation facility, then transmitted through high-voltage wires to central load areas. After that, the voltage is reduced before being distributed to individual consumers.

The conventional grid operates through one-way communication as electricity generated at power plants is transmitted to consumers with minimal interaction or feedback from end-users [6]. Some cons are inadequate power distribution, extremely high operational costs, significant problems in deploying renewable energy sources, and, very importantly, the high failure rate. Conventional grids are also prone to outages, lacking real-time monitoring and self-healing properties [7].

A smart grid is a modern electricity distribution network that utilizes computers and other electronic tools to collect information, including power consumption data, send it across the entire system, and then act on this interactivity between suppliers and users in real-time. The represents a significant improvement over the traditional, one-way delivery of electricity favored by power plants to consumers, enabling two-way communication and active control features [8].

# CHARACTERISTICS OF SMART GRIDS

Smart grids are featured by various characteristics, as listed in Fig. (**1**). For instance, in self-healing, these grids can identify and respond to faults on their own, thus being dependable and reducing their downtimes. This also encourages active customer participation in the management of energy usage, generally through demand response programs. Moreover, grids can resist both physical and cyber attacks. Moreover, smart grids have the capacity to recover from such attacks and guarantee a stable power supply [9].

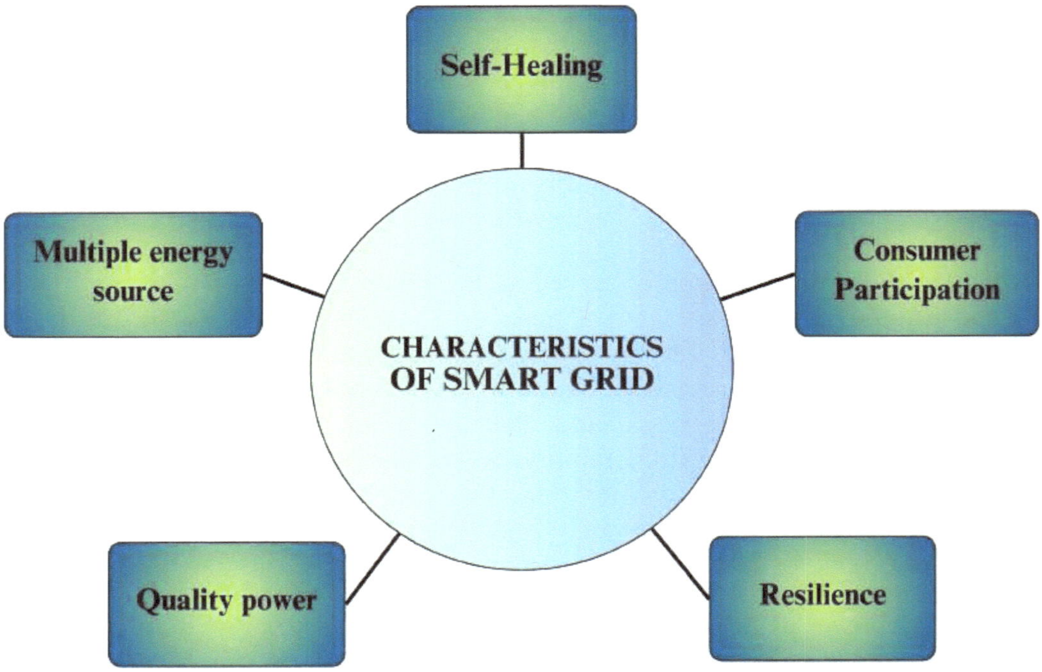


**Fig. (1).** Key characteristics of smart grid systems.

Also, smart grids are appropriate for modern users, as these grids provide high-quality power, minimizing issues like voltage sags and surges. Additionally, smart grids can interconnect, enable, and manage multiple generation options, including renewable resources or distributed generators, as shown in Fig. (**2**).

The core of a smart grid is the integration of green power sources, such as solar panels on our roof, with conventional grid systems [10]. These renewable energy sources are complementary to conventional power generation and help reduce the carbon footprint [11]. Plug-in hybrid electric vehicles, as shown in Fig. (**2**), also become part of this smart grid system. Most of these vehicles are capable of consuming electricity from the grid and also returning power back during times when the system's demand is high [12]. Also, smart thermostats, appliances, and in-home control devices can utilize the power of this system. The real-time pricing signals allow consumers to track and manage their energy usage (energy efficiency) in order to make cost savings [13].

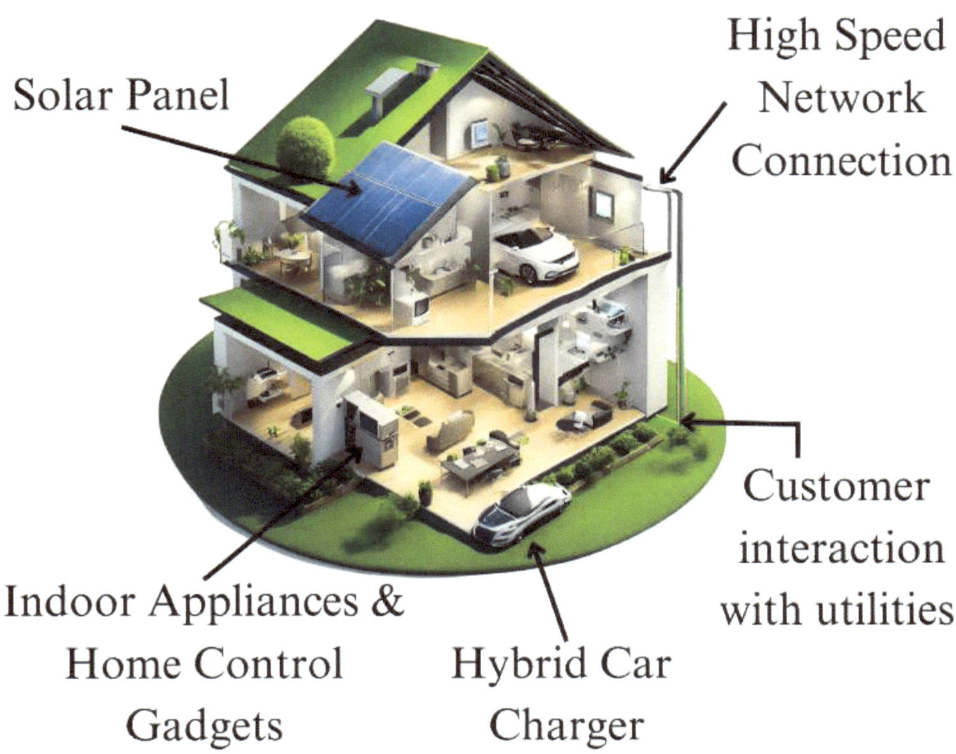


**Fig. (2).** Smart home energy management system.

By connecting homes to a high-speed, networked connection to utilities, the infrastructure would enable two-way communication that ultimately allows continuous operation on an ongoing basis for Virtual Power Plants (VPPs) and Distributed Energy Resource Management Systems [14].

Smart grid systems communicate through control centres, which are essentially nerve centres of these grids that help in managing communication and electricity flow throughout the system. The control centre is responsible for the operation of the entire network, while the distribution control center, on the other hand, manages activities that concern distribution, as shown in Fig. (**3**) [15]. The energy service provider is the main source of electricity that comes from different places like power plants built or generation stations, as shown in Fig. (**3**) [16].

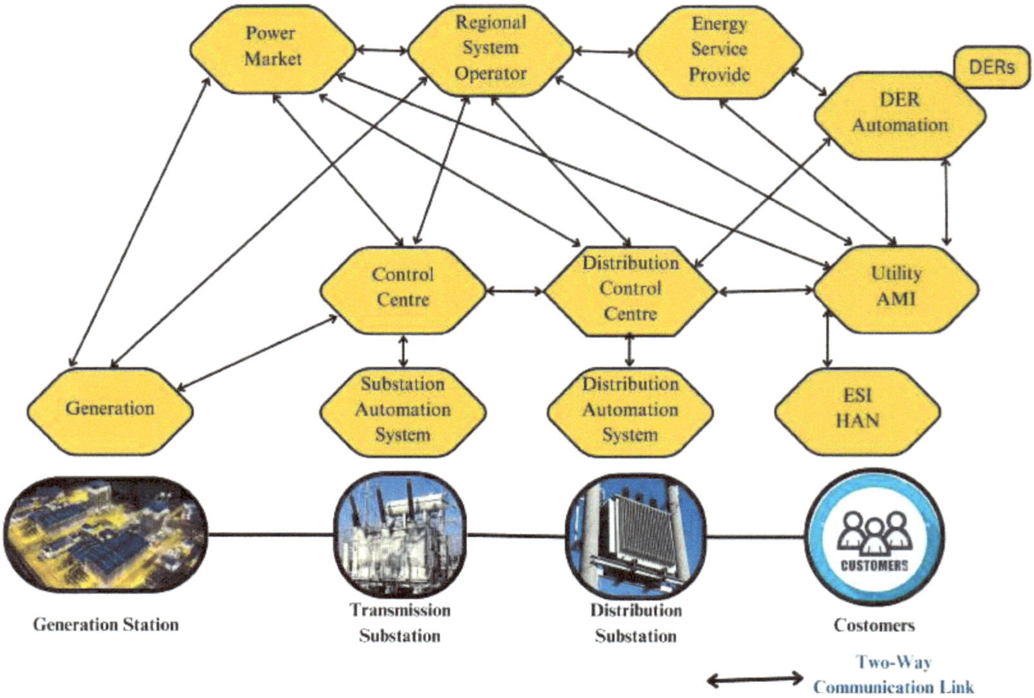


**Fig. (3).** Smart Grid Communication and Control Infrastructure.

The smart grid system allows electricity to flow from various generation sources to diverse types of consumers. Fig. (**4**) illustrates the various sources contributing to the smart grid on one side and includes traditional power generation sources, such as nuclear and thermal power plants, and factories, along with other renewable energy sources such as hydropower (*e.g.*, a hydroelectric dam), wind turbines, and photovoltaic.

The regional system operator is important for balancing supply and demand on a larger scale, resulting in reliable power delivery more efficiently on the main street. The power market enables electricity trading and contributes to the integration of Distributed Energy Resources [17].

Transmission and Distribution Substations are an important part of electrical generation, distribution systems, respectively. It is the substation where electricity is used to step up or down through a voltage converter transformer for transmission and distribution over long distances [18]. Residential homes and cities/offices are highlighted as the biggest electricity consumers of smart grid systems. Real-time monitoring, control and optimization of the smart grid operations are facilitated by automation systems at different levels (such as substation automation, distribution automation and DER Automation). Indeed, two-way communication between customers and the utility is a cornerstone of the smart grid to enable demand response programs, dynamic pricing options, and customer engagement in energy management. The smart grid ensures more efficient distribution and utilization of electricity, as well as the development towards a sustainable energy system using distributed and low-carbon sources for central generation, while also stimulating the end-user to take responsibility for their own power use.

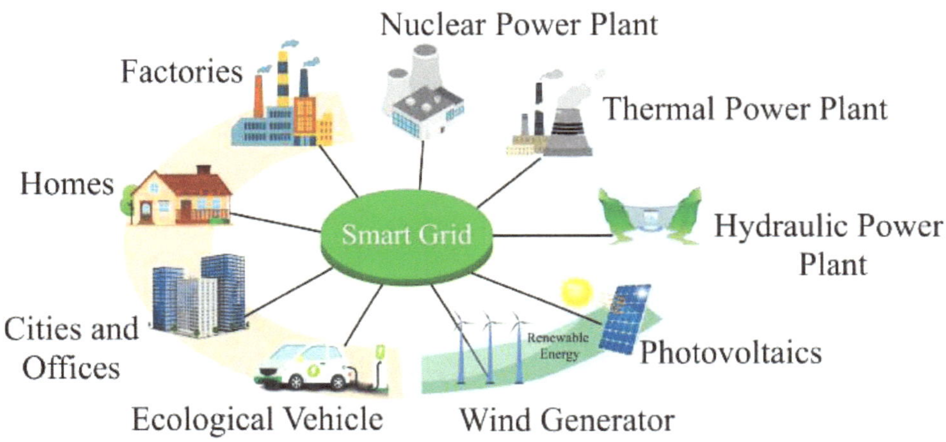


**Fig. (4).** Smart grid ecosystem and its components.

## KEY COMPONENTS OF SMART GRID SYSTEM

The Smart Grid system represents a transformative approach to modernizing power systems by integrating advanced technologies to enhance efficiency, reliability, and sustainability. Various network topologies, such as NAN, SAN, FAN, and WSN, play an important role in enabling seamless communication across different components of the grid. Additionally, smart grid technologies such as the IoT, machine learning, artificial intelligence, big data analytics, and neural networks contribute to optimizing operations. Innovative methods like short-term memory networks and power line communication technologies further enhance the adaptability and intelligence of the grid [19]. The classification of different key components used in the smart grid system is listed in Table **1**.

Data security and encryption are critical aspects of the Smart Grid, with strategies such as multidimensional data aggregation and cognitive risk control ensuring the secure transmission of information. The grid accommodates both AC and DC current transmission, making it versatile for various infrastructure needs. Key applications include advanced metering infrastructure, substation and distribution automation, and strategic planning for transmission and production. Connectivity is bolstered by technologies such as Ethernet, Wi-Fi, sensors, and emerging 5G networks, offering high-speed and reliable communication. Furthermore, analytical tools like time series analysis and regression models are utilized for effective decision-making, enabling energy providers to predict trends and improve system performance [20].

**Table 1. Classification of different key components used in Smart Grid System [19].**

| Sr No | Classifications | Examples |
|---|---|---|
| 1 | Network Topology | NAN (Neighbourhood Area Network), SAN (Storage Area Network), FAN (Field Area Network), and WSN (Wireless Sensor Network). |
| 2 | Smart Grid Technology | Internet of Things (IoT), Machine learning and artificial intelligence, big data analytics, Neural networks, Short-term memory network, and Power Line Communication Technology. |
| 3 | Encryption | Multidimensional Data Aggregation, and Cognitive Risk Control. |
| 4 | Current and Data Transmission | Yes and No options (referring to AC and DC transmission). |
| 5 | Applications | Advanced Metering Infrastructure, Substation Automation, Distribution Automation, Distributed Energy Resources, Transmission Planning, and Production Planning. |
| 6 | Connectivity | Ethernet, Wi-Fi, Sensors, and Emphasis on 5G connectivity. |
| 7 | Tools | Time Series Analysis and Regression Model. |

## AI APPROACHES IN SMART GRID SYSTEMS

Different AI approaches have been used for the advancement of smart grid systems, such as Long Short Term Memory - Reinforcement Learning, Hybrid Convolutional Neural Network, Neural Networks, Random Forest, Convolutional neural networks, Autoencoders, and Deep Q-Networks, as shown in Fig. (**5**).

Long Short Term Memory - Reinforcement Learning(LSTM-RL) is the combination of Long Short Term Memory (a deep learning design uniquely created to deal with sequential information) with Reinforcement Learning (a subfield in AI taking care of decision making). LSTM models are utilized to control the progression of data through gates that hold historical data in a sequential manner, especially with terrestrial data. Reinforcement learning addresses these difficulties by interacting with the environment, getting feedback, and acquiring rewards. The combination of LSTM and reinforcement learning in LSTM-RL merges the predictive capabilities of LSTM with the advancement of decision-making reinforcement learning, especially in tasks like demand response and energy forecasting. For instance, in a study by Mocanu et al, LSTM-RL was used for real-time energy optimization in buildings, demonstrating its ability to manage energy consumption while maintaining occupant comfort adaptively. This model enables accurate energy demand prediction and adaptive utilization strategies. The LSTM component figures out how much energy the building will require in the future based on the observed activity patterns, and Reinforcement Learning resolves a strategy to balance demand with supply, considering costs as well as more worldwide grid stability considerations [21] [22].

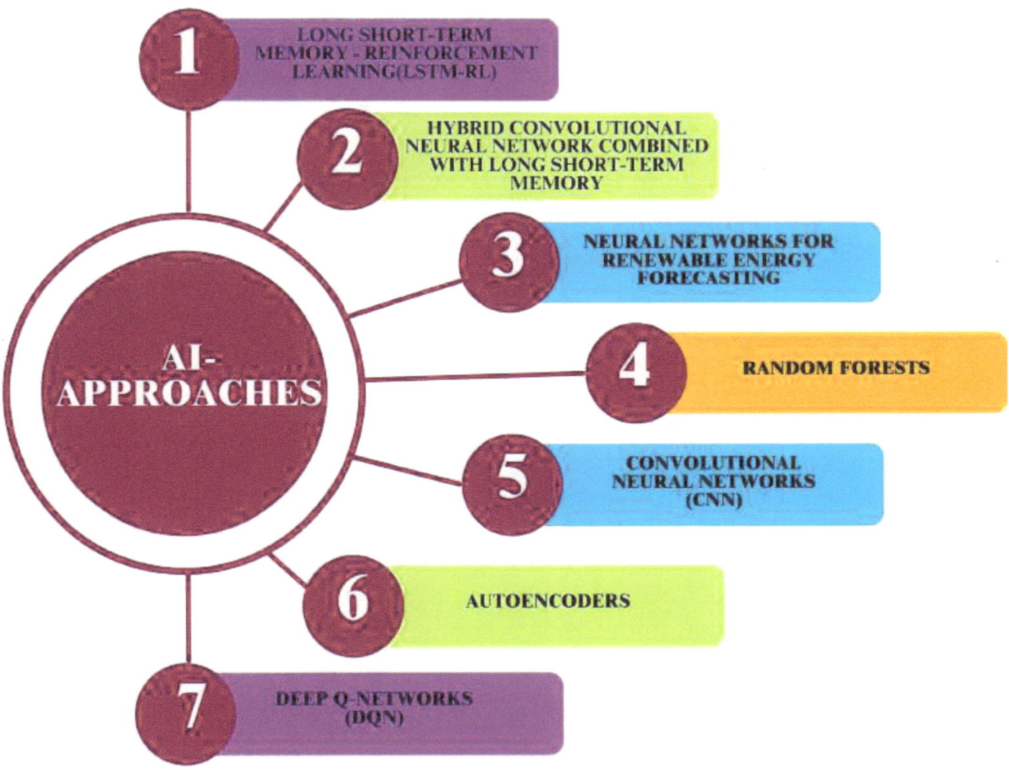


**Fig. (5).** AI Approaches for Smart Grid Systems

The Hybrid Convolutional Neural Network, combined with Long Short-Term Memory (hybrid CNN-LSTM), model enjoys the benefits of utilizing the spatial features provided by CNN, and terrestrial conditions are effortlessly captured due to LSTM. CNNs are proficient at handling multi-dimensional input data for energy forecasting, for instance, consolidating weather conditions or geological data as elements. These elements are then used to make predictions utilizing LSTMs that capture information throughout a period of time. This engineering is especially useful for overseeing complex, multivariate energy scenes impacted by terrestrial or spatial elements. The CNN part lessens the dimensionality of input information and distinguishes significant patterns, while the LSTM parts handle sequence learning related to energy utilization over time. For example, Kim and Cho applied a CNN-LSTM model to predict household energy consumption patterns in South Korea. Their model effectively learned both spatial and temporal dependencies, resulting in highly accurate predictions [23] [24].

Neural Networks, especially deep learning models, play a critical part in renewable energy forecasting. These models excel at capturing unpredictable, non-linear relationships among various elements that impact renewable energy generation, such as weather patterns, seasonal variations, and hardware productivity. Different neural network architectures like Multi-layer Perceptrons (MLPs), Recurrent Neural Networks (RNNs), and Long Short Term Memory (LSTM) are commonly used in this field. By analyzing vast amounts of historical data, meteorological figures, and real-time measurements, these models can predict renewable energy production across various time scales, going from the present moment to long-term gauges. The precise forecast of an environmentally friendly power age helps matrix administrators manage the discontinuity of sustainable sources and upgrade framework activities to ensure productivity. For instance, artificial neural networks were implemented for demand-side management in Brazil's smart grid environment. The model analyzed real-time data and helped optimize energy usage, demonstrating how neural networks can reduce peak loads and improve grid stability [25] [26].

Random Forests are an ensemble learning technique that aggregates multiple decision trees to predict the outcome more accurately and reliably. In asset management and predictive maintenance applications, Random Forests are found to be very effective in processing a vast variety of input features, including the age of equipment, operational conditions, maintenance history, and sensor data. Each decision tree in the Random Forest is trained on a random subset of both data and features; thus it prevents overfitting, which enhances the model to generalize well to new data. Random Forests work very well with both categorical and numerical data, so they can fit various maintenance-related tasks. These can predict the timing of failures, estimate the remaining useful life of asset stock based on current condition, and determine key factors driving asset degradation. In addition to this, Random Forests return ranks of feature importance that explain those variables most responsible for decisions made in maintaining turbines. For example, a Random Forest-based model was developed for fault prediction in China's smart distribution networks. Their model accurately predicted faults using sensor and operational data, helping to reduce downtime and improve maintenance scheduling [27] [28].

Convolutional Neural Networks (CNNs), originally created to process images, have begun to be used in asset management and predictive maintenance, particularly in the context of processing complex, multidimensional data. CNNs can successfully treat time series data or spectral measurements acquired from multiple sensors in power systems as a 2D or 3D representation of such data. This way, the network can learn the spatial and temporal dependencies of the equipment's behaviour. In CNNs, convolutional layers are used to extract important features from the input data, pooling layers are used to reduce dimensionality, and fully connected neurons in the last layer are used for generating predictions based on the learned features. They are great for fault classification, anomaly detection in power quality data or predicting equipment failures from image/signal-based inspection. For example, wide and deep CNNs were applied for electricity-theft detection in China's smart grid systems. Their model analyzed consumption patterns and visualized meter data to identify anomalies, significantly improving theft detection accuracy [29] [30].

Autoencoders belong to a set of unsupervised learning models that can take input data, compress it into a smaller size, and then reconstruct it as an output, essentially serving as a decoder. Auto encoders are used extensively to recover normal operation for fault detection and anomaly detection in power systems. Autoencoders are very effective for anomaly detection, as amplification of the reconstruction error indicates a potential fault. Variants such as denoising autoencoders and variational autoencoders enhance the robustness and generalizability of the anomaly detection process. Autoencoders reduce high-dimensional sensor data to a set of essential features that capture typical system operations. These are well-suited for determining minor aberrations in typical grid operations, including anomaly voltage or current readings that can represent early warnings to equipment failure or potential cyber attacks on the power system. Sakurada and Yairi used autoencoders to detect anomalies in time-series sensor data from power systems. The model flagged abnormal behavior based on high reconstruction errors, making it effective for early fault detection in smart grid environments [31] [32].

Deep Q-Networks employ a combination of deep learning and Q-learning, a method of reinforcement learning, for solving complex decision-making tasks in grid operations and control. DQNs are modifications of neural networks specifically built to approximate the Q-function. The Q-function estimates how desirable it is to take particular actions in a certain state. Such networks enable the development of successful control strategies for large state spaces that are commonly associated with power systems. For example, DQNs may be used to regulate voltages in grid operations, supervise energy storage devices, or control demand response programs. Through interactions either with a simulated or physical environment, the algorithm acquires the ability to make decisions that maximize long-run rewards such as stability and cost savings in grids. For instance, a Deep Q-Network to coordinate electric vehicle charging across multiple agents in a smart grid was implemented by Maguluri et al. The system reduced peak demand and balanced grid load effectively, showing the potential of DQNs in managing decentralized energy resources [33] [34].

The integration of AI techniques in Smart Grids has enhanced reliability, efficiency, and operational optimization. However, this integration brings both benefits and challenges, as shown in Fig 6.

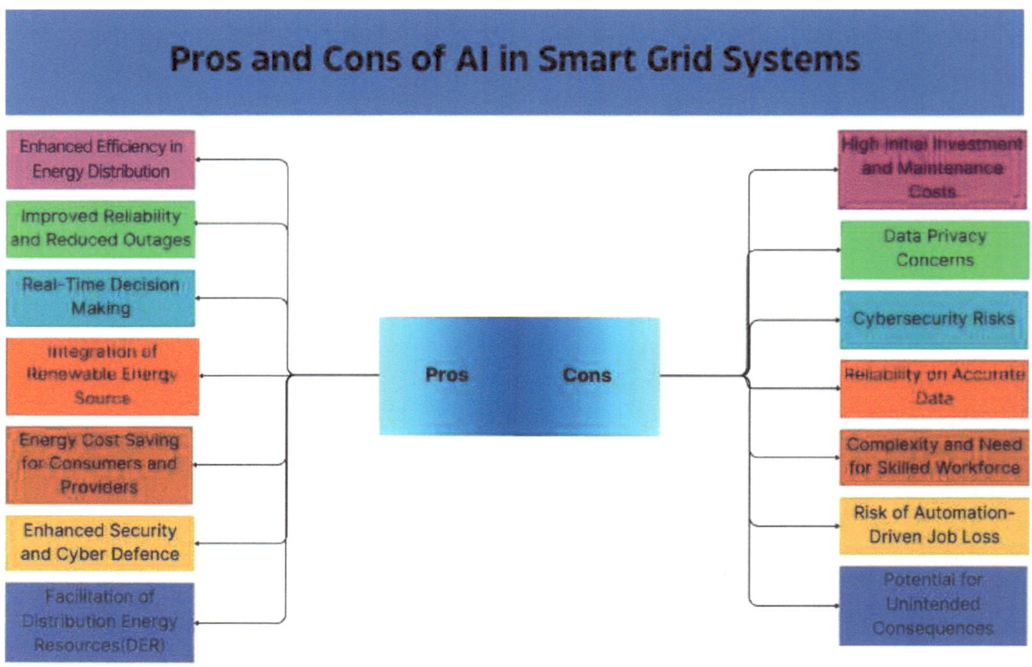


**Fig. (6).** Pros and Cons of AI in Smart Grid Systems.

## CHALLENGES IN THE IMPLEMENTATION OF AI IN SMART GRIDS

The implementation of AI in smart grid systems has various problems to be addressed to ensure its sustainability and success. Data security and privacy become major concerns because of the high presence of sensitive consumer data in the widespread use of AI, thereby requiring strong measures to avoid breaches and maintain privacy with set regulatory requirements [35]. Besides, an AI-based grid inherently presents cybersecurity threats, with the risk of advanced attacks to potentially interrupt operations or manipulate energy distribution [36].

The performance of AI output is highly dependent on data quality and availability, but inefficiencies occur due to inconsistent or corrupt input from sensors and older systems. Integration is another issue because legacy systems may be the oldest system and, normally, their compatibility must be highly upgraded for continued use with an added high cost. Scalability presents some of the biggest challenges (Fig. **7**) because the AI models must be customized and tailored heavily for effective operation on high-power, intricate grids [37]. Furthermore, a lack of people with AI and data science, as well as power engineering skills, affects efficient implementation, and strong regulatory and compliance requirements may elevate the complexity of deploying those systems [38].

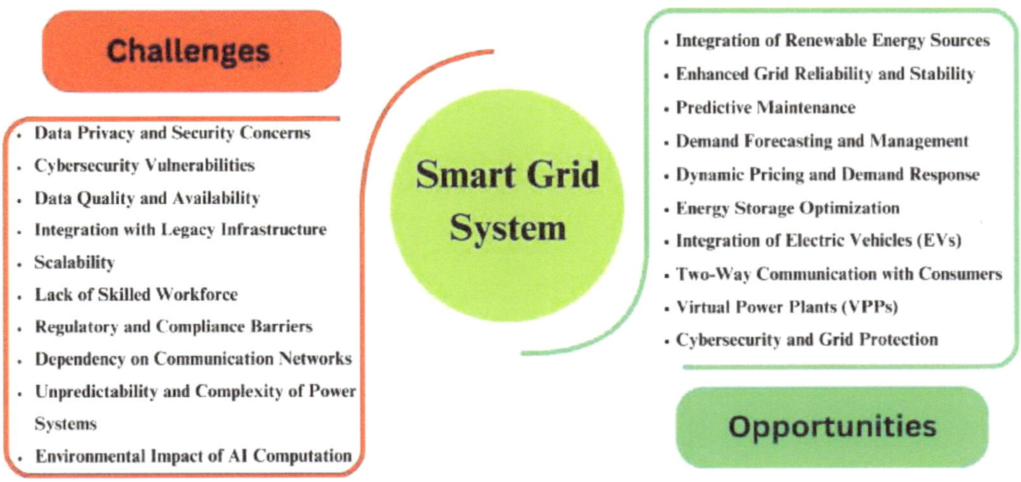


**Fig. (7).** Challenges and Opportunities in Smart Grid Systems.

Also, reliance on communication networks introduces potential points of failure since outages can impact operations at grid levels [39]. The inherent uncertainty of power systems, where demand cannot be predicted and will vary over time, also poses a challenge to reliability for AI. Yet the energy intensity associated with large-scale computations in AI means that environmental gains from smart grids can be unfairly nullified without appropriate management.

## OPPORTUNITIES IN THE ADVANCEMENT OF SMART GRID SYSTEMS

Smart grid systems incorporating artificial intelligence are transformative and offer various opportunities for enhancing energy efficiency, reliability, and sustainability. The integration of such renewable sources of energy as solar and wind energy can be assisted with ease with a prediction of the pattern of generation against the demand-supply balance using AI. It further enhances a high level of reliability of the grid through real-time monitoring, anomaly detection, and automatic responses that stabilize the voltage and prevent outages from arising [40]. AI-driven demand forecasting and dynamic pricing optimize energy distribution and encourage off-peak usage, thus easing the pressure on the grid. AI transforms predictive maintenance by analyzing the behaviour of equipment and predicting possible failures; both time and costs are reduced. AI helps optimize energy storage, smart electric vehicle charging, and vehicle-to-grid integration, that is, in fact, EVs themselves become distributed energy resources. AI further enhances the pliability by leveraging VPP and two-way communication capabilities with consumers, while strong cybersecurity will protect the critical infrastructure from probable threats.

## CONCLUSION

The move from traditional power grids to smart grids highlights a significant progression for the distribution and management of electricity. Smart grids that leverage digital advancements and notable two-way communication avenues offer efficiency, reliability, and sustainability, which are evident today. Advancements in artificial intelligence increase the effectiveness of smart grid solutions, creating optimization, predictive maintenance, and adaptive control avenues. Throughout this chapter, we have discussed several components of a smart grid and some AI approaches that can be applied in making the grids smarter. These measures include Demand Response, forecasting, and optimizing asset loading, among others, which AI has the potential to revolutionize the production, distribution, an consumption of electricity. Nonetheless, the full embracing of AI in the smart grid faces a myriad of risks. Digital systems present continuous threats from malicious software that can cause loss of data and interception in the systems. The digital transformation amplifies the risk due to the interconnectedness of components, necessitating more robust security measures. Human-made risks, both accidental and intentional, remain a big threat to the proper functioning of the integration. The threats necessitate the development of systems that incorporate stringent security measures, behavioral monitoring, and training. The socioeconomic impact is a significant inhibitor of the full adoption of smart grid technology. Stakeholders may fear data access challenges due to the initial costs of developing such technology. The challenges will need close cooperation, and standardization integration efforts of all stakeholders for sustainable business practices. However, despite the contested theoretical hurdles, the benefits of integrating AI into the smart grid are measurable and worthwhile. Improved energy management, greater efficiency, and better integration of renewable sources can be achieved. Further, AI outputs help prolong the lifespan of assets, reduce outages frequency from grid disturbances, and promise less downtime. Moreover, with the real possibility of massive adoption of complex power systems running with more unstable energy inputs, the smart grid can continue to supply power and predictive maintenance can significantly minimize the negative impacts. AI offers improved grid stability, reduces the chance of system failures relatively faster than a grid operator. For the future of smarter grids, smart grids with AI will continue to develop with the times. Smart grids made smarter by AI will become more powerful due to the learning the latter acquires and become more efficient by increasing social, economic, and technical competence. Thus, despite the downsides, AI-integrated smart grids have a bright future in the energy ecosystem where intelligence is well appreciated.

## REFERENCES

[1] S. Scherbov, W. Lutz, and W.C. Sanderson, "The uncertain timing of reaching 8 billion, peak world population, and other demographic milestones", *Popul. Dev. Rev.,* vol. 37, no. 3, pp. 571-578, 2011. [http://dx.doi.org/10.1111/j.1728-4457.2011.00435.x] [PMID: 22167816]

[2] T.M. Gür, "Giga-ton and tera-watt scale challenges at the energy - climate crossroads: A global perspective", *Energy,* vol. 290, p. 129971, 2024. [http://dx.doi.org/10.1016/j.energy.2023.129971]

[3] E.K. Schwartz, and M. Krarti, "Review of adoption status of sustainable energy technologies in the US residential building sector", *Energies,* vol. 15, no. 6, p. 2027, 2022. [http://dx.doi.org/10.3390/en15062027]

[4] F.F. Wu, K. Moslehi, and A. Bose, "Power system control centers: Past, present, and future", *Proc. IEEE,* vol. 93, no. 11, pp. 1890-1908, 2005. [http://dx.doi.org/10.1109/JPROC.2005.857499]

[5] Imdadullah, B. Alamri, M.A. Hossain, and M.S.J. Asghar, "Electric power network interconnection: A review on current status, future prospects and research direction", *Electronics (Basel),* vol. 10, no. 17, p. 2179, 2021. [http://dx.doi.org/10.3390/electronics10172179]

[6] C.H. Lo, and N. Ansari, "The progressive smart grid system from both power and communications aspects", *IEEE Commun. Surv. Tutor.,* vol. 14, no. 3, pp. 799-821, 2011. [http://dx.doi.org/10.1109/SURV.2011.072811.00089]

[7] P.K. Maurya, "Self-healing grids: AI techniques for automatic restoration after outages", *Power System Technology,* vol. 48, no. 1, pp. 494-510, 2024. [http://dx.doi.org/10.52783/pst.302]

[8] V.C. Gungor, D. Sahin, T. Kocak, S. Ergut, C. Buccella, C. Cecati, and G.P. Hancke, "Smart grid technologies: Communication technologies and standards", *IEEE Trans. Industr. Inform.,* vol. 7, no. 4, pp. 529-539, 2011. [http://dx.doi.org/10.1109/TII.2011.2166794]

[9] S. Paul, M.S. Rabbani, R.K. Kundu, and S.M.R. Zaman, "A review of smart technology (Smart Grid) and its features", [http://dx.doi.org/10.1109/ICONCE.2014.6808719]

[10] A.T. Hoang, V.V. Pham, and X.P. Nguyen, "Integrating renewable sources into energy system for smart city as a sagacious strategy towards clean and sustainable process", *J. Clean. Prod.,* vol. 305, p. 127161, 2021. [http://dx.doi.org/10.1016/j.jclepro.2021.127161]

[11] T. Kataray, B. Nitesh, B. Yarram, S. Sinha, E. Cuce, S. Shaik, P. Vigneshwaran, and A. Roy, "Integration of smart grid with renewable energy sources: Opportunities and challenges – A comprehensive review", *Sustain. Energy Technol. Assess.,* vol. 58, p. 103363, 2023. [http://dx.doi.org/10.1016/j.seta.2023.103363]

[12] R.A. Waraich, M.D. Galus, C. Dobler, M. Balmer, G. Andersson, and K.W. Axhausen, "Plug-in hybrid electric vehicles and smart grids: Investigations based on a microsimulation", *Transp. Res., Part C Emerg. Technol.,* vol. 28, pp. 74-86, 2013. [http://dx.doi.org/10.1016/j.trc.2012.10.011]

[13] S.J. Darby, "Demand response and smart technology in theory and practice: Customer experiences and system actors", *Energy Policy,* vol. 143, p. 111573, 2020. [http://dx.doi.org/10.1016/j.enpol.2020.111573]

[14] G. Dileep, "A survey on smart grid technologies and applications", *Renew. Energy,* vol. 146, pp. 2589-2625, 2020. [http://dx.doi.org/10.1016/j.renene.2019.08.092]

[15] G. Suter, and T.G. Werner, "The distribution control centre in a smartgrid", *20th Int. Conf. and Exhb. on Elect. Dist.,* CIRED 2009. [http://dx.doi.org/10.1049/cp.2009.0521]

[16] S. Karnouskos, P.G. Da Silva, and D. Ilic, "Energy services for the smart grid city", *2012 6th IEEE International Conference on Digital Ecosystems and Technologies (DEST),* pp. 1-6, . Campione d'Italia, Italy [http://dx.doi.org/10.1109/DEST.2012.6227925]

[17] C. Cecati, C. Citro, and P. Siano, "Combined operations of renewable energy systems and responsive demand in a smart grid", *IEEE Trans. Sustain. Energy,* vol. 2, no. 4, pp. 468-476, 2011. [http://dx.doi.org/10.1109/TSTE.2011.2161624]

[18] V. Hamidi, K.S. Smith, and R.C. Wilson, "Smart grid technology review within the transmission and distribution sector, in 2010 IEEE PES Innovative Smart Grid Technologies Conference Europe", *ISGT Europe,* pp. 1-8, 2010.

[19] J.J. Moreno Escobar, O. Morales Matamoros, R. Tejeida Padilla, I. Lina Reyes, and H. Quintana Espinosa, "A comprehensive review on smart grids: Challenges and opportunities", *Sensors (Basel),* vol. 21, no. 21, p. 6978, 2021. [http://dx.doi.org/10.3390/s21216978] [PMID: 34770285]

[20] P.K. Pandey, R. Kumar, V. Gupta, A. Kanungo, and S. Diwania, "An innovative hybrid controller-based combined grid-connected hybrid renewable energy system", *Electr. Eng.,* vol. 106, no. 6, pp. 7055-7072, 2024. [http://dx.doi.org/10.1007/s00202-024-02363-2]

[21] E. Mocanu, D.C. Mocanu, P.H. Nguyen, A. Liotta, M.E. Webber, M. Gibescu, and J.G. Slootweg, "On-line building energy optimization using deep reinforcement learning", *IEEE Trans. Smart Grid,* vol. 10, no. 4, pp. 3698-3708, 2019. [http://dx.doi.org/10.1109/TSG.2018.2834219]

[22] X. Zhou, W. Lin, R. Kumar, P. Cui, and Z. Ma, "A data-driven strategy using long short term memory models and reinforcement learning to predict building electricity consumption", *Appl. Energy,* vol. 306, p. 118078, 2022. [http://dx.doi.org/10.1016/j.apenergy.2021.118078]

[23] T.Y. Kim, and S.B. Cho, "Predicting residential energy consumption using CNN-LSTM neural networks", *Energy,* vol. 182, pp. 72-81, 2019. [http://dx.doi.org/10.1016/j.energy.2019.05.230]

[24] P. Ding, J. Li, L. Wang, M. Wen, and Y. Guan, "HYBRID□CNN: An efficient scheme for abnormal flow detection in the SDN-based smart grid", *Secur. Commun. Netw.,* vol. 2020, no. 1, pp. 1-20, 2020.

[http://dx.doi.org/10.1155/2020/8850550]

[25] P.K. Pandey, Z. El Khatib, R. Kumar, M. Singla, M. Safaraliev, and F. Kamalov, "A novel hybrid control strategy for grid-connected pv-battery system under different operating conditions", *Przegl. Elektrotech.,* vol. 9, 2024..

[26] M.N. Macedo, J.J. Galo, L.A.L. De Almeida, and A.D.C. Lima, "Demand side management using artificial neural networks in a smart grid environment", *Renew. Sustain. Energy Rev.,* vol. 41, pp. 128-133, 2015.
[http://dx.doi.org/10.1016/j.rser.2014.08.035]

[27] M. Cerrada, R-V. Sánchez, C. Li, F. Pacheco, D. Cabrera, J. Valente de Oliveira, and R.E. Vásquez, "A review on data-driven fault severity assessment in rolling bearings", *Mech. Syst. Signal Process.,* vol. 99, pp. 169-196, 2018.
[http://dx.doi.org/10.1016/j.ymssp.2017.06.012]

[28] R. Lin, Z. Pei, Z. Ye, B. Wu, and G. Yang, "A voted based random forests algorithm for smart grid distribution network faults prediction", *Enterprise Inf. Syst.,* vol. 14, no. 4, pp. 496-514, 2020.
[http://dx.doi.org/10.1080/17517575.2019.1600724]

[29] H. Kaplan, K. Tehrani, and M. Jamshidi, "Fault diagnosis of smart grids based on deep learning approach", *2021 World Automation Congress (WAC),* pp. 164-169, 2021.
[http://dx.doi.org/10.23919/WAC50355.2021.9559474]

[30] Z. Zheng, Y. Yang, X. Niu, H.N. Dai, and Y. Zhou, "Wide and deep convolutional neural networks for electricity-theft detection to secure smart grids", *IEEE Trans. Industr. Inform.,* vol. 14, no. 4, pp. 1606-1615, 2018.
[http://dx.doi.org/10.1109/TII.2017.2785963]

[31] M. Sakurada, and T. Yairi, "Anomaly detection using autoencoders with nonlinear dimensionality reduction", *Proceedings of the MLSDA 2014 2nd workshop on machine learning for sensory data analysis,* pp. 4-11, 2014.
[http://dx.doi.org/10.1145/2689746.2689747]

[32] A. Zafar, Y. Che, M. Faheem, M. Abubakar, S. Ali, and M.S. Bhutta, "Machine learning autoencoder-based parameters prediction for solar power generation systems in smart grid", *IET Smart Grid,* vol. 7, no. 3, pp. 328-350, 2024.
[http://dx.doi.org/10.1049/stg2.12153]

[33] C. Guan, Y. Wang, X. Lin, S. Nazarian, and M. Pedram, "Reinforcement learning-based control of residential energy storage systems for electric bill minimization",

[34] L.P. Maguluri, A. Umasankar, D. Vijendra Babu, A.S. Anselin Nisha, M.R. Prabhu, and S.A. Tilwani, "Coordinating electric vehicle charging with multiagent deep Q-networks for smart grid load balancing", *Sustain. Comput.,* vol. 43, p. 100993, 2024.
[http://dx.doi.org/10.1016/j.suscom.2024.100993]

[35] P. McDaniel, and S. McLaughlin, "Security and privacy challenges in the smart grid", *IEEE Secur. Priv.,* vol. 7, no. 3, pp. 75-77, 2009.
[http://dx.doi.org/10.1109/MSP.2009.76]

[36] Y. Yang, T. Littler, S. Sezer, K. McLaughlin, and H.F. Wang, "Impact of cyber-security issues on smart grid", *Proc. IEEE PES Int. Conf. Exhib. Innov. Smart Grid Technol. (ISGT),* pp. 1-7, 2011.

[37] K. May, P. Vingerhoets, and L. Sigrist, "Barriers regarding scalability and replicability of smart grid projects", *Proc. Int. Conf. Eur. Energy Market (EEM),* pp. 1-5, 2015.

[38] Y. Yan, Y. Qian, H. Sharif, and D. Tipper, "A survey on smart grid communication infrastructures: Motivations, requirements and challenges", *IEEE Commun. Surv. Tutor.,* vol. 15, no. 1, pp. 5-20, 2013.
[http://dx.doi.org/10.1109/SURV.2012.021312.00034]

[39] R. Ma, H.H. Chen, Y.R. Huang, and W. Meng, "Smart grid communication: Its challenges and

opportunities", *IEEE Trans. Smart Grid,* vol. 4, no. 1, pp. 36-46, 2013. [http://dx.doi.org/10.1109/TSG.2012.2225851]

[40] N. Phuangpornpitak, and S. Tia, "Opportunities and challenges of integrating renewable energy in smart grid system", *Energy Procedia,* vol. 34, pp. 282-290, 2013. [http://dx.doi.org/10.1016/j.egypro.2013.06.756]

# CHAPTER 2

# Renewable Energy Innovations: A Path to Sustainability

**Tanya Sood**[1], **Raman Kumar**[2], **Abhishek Kandwal**[3], **Sushil Kumar**[4] and **Pawan Kumar**[1,*]

[1] *School of Physics and Materials Science, Shoolini University, Solan 173229, Himachal Pradesh, India*

[2] *Department of Physics, Government College Daulatpur Chowk, Una 177204, Himachal Pradesh, India*

[3] *School of Chips, Xi'an Jiaotong-Liverpool University, Taicang 215400, Suzhou, China*

[4] *School of Core Engineering, Shoolini University, Solan 173229, Himachal Pradesh, India*

**Abstract:** This chapter offers an in-depth exploration of different renewable energy sources that play a crucial role in advancing sustainable global energy systems. It thoroughly explores five primary renewable technologies: solar energy, wind energy, geothermal energy, hydro energy, and biomass energy, delving into the fundamental principles underlying each technology and detailing how they convert natural resources into usable energy. The analysis evaluates the strengths and weaknesses of each technology, focusing on their efficiency, environmental impacts, and economic viability. Furthermore, the chapter reviews the current advancements and successful implementations of these technologies across different regions, showcasing practical examples of their deployment. It also anticipates future applications by examining emerging trends, technological innovations, and potential advancements that can further solidify their role in meeting sustainable energy targets. This chapter provides a detailed overview and offers readers a comprehensive analysis of how renewable energy technologies are essential in addressing the global energy crisis and promoting a sustainable future. It underscores the critical need for continued innovation and strategic planning to optimize the integration and management of these technologies, ultimately highlighting their potential to revolutionize the energy sector and support a cleaner, more resilient world.

**Keywords:** Energy efficiency, Environmental impact, Renewable energy technologies, Sustainable energy systems, Technological innovation.

* **Corresponding author Pawan Kumar:** School of Physics and Materials Science, Shoolini University, Solan 173229, Himachal Pradesh, India; E-mail: kumar74pawan@gmail.com

**Reetu Malhotra, Pooja Mahajan, Manish Singla & Ramesh Kumar (Eds.)**

# INTRODUCTION

## Importance of Renewable Energy in Sustainable Systems

The global energy sector is experiencing a significant shift, propelled by the urgent need to transition from fossil fuels to more sustainable and environmentally friendly energy alternatives [1 - 5].Traditional energy systems, largely based on coal, oil, and natural gas, have been instrumental in fuelling industrialization and economic growth.However, the environmental consequences of these fossil fuel-based systems, particularly their contribution to climate change, air pollution, and ecosystem degradation, have become increasingly evident. As the world faces the combined problem of fulfilling expanding energy demand while reducing environmental effects, renewable energy has emerged as a critical alternative [6, 7]. Renewable energy is defined as energy generated from naturally occurring processes that are continuously replenished. These energy sources, such as sunlight, wind, water, biomass, and geothermal heat, offer a limitless and eco-friendly substitute for fossil fuels. Unlike limited fossil fuel resources, renewable energy sources are naturally regenerative and produce negligible to no greenhouse gases during operation, making them critical elements of sustainable systems. Renewable energy is important not just for the environment [8, 9]. It plays a key role in promoting energy security, reducing dependence on imported fuels, and promoting economic development [10]. As renewable technologies advance, their incorporation into energy systems may lead to more robust, decentralized power grids and the development of new industries and jobs. Furthermore, renewable energy technologies can overcome energy access issues, especially in isolated or underserved locations, through off-grid or distributed energy solutions [11].

In the current context of climate change, the move to renewable energy is no longer a choice, but an imperative. The Paris Agreement and numerous national commitments underscore the need to limit global warming to well below 2°C above pre-industrial levels, with efforts to keep it at 1.5°C. Achieving these targets will require a massive scale-up of renewable energy technologies, coupled with energy efficiency measures, to decarbonize power generation, transportation, and industry.

## Overview of Key Renewable Technologies

This chapter explores five key renewable energy technologies, as shown in Fig. (**1**), that are transforming the global energy landscape: solar energy, wind energy, geothermal energy, hydro energy, and biomass energy [12, 13]. Each of these technologies leverages different natural resources to generate power, resulting in a more diverse and robust energy mix.

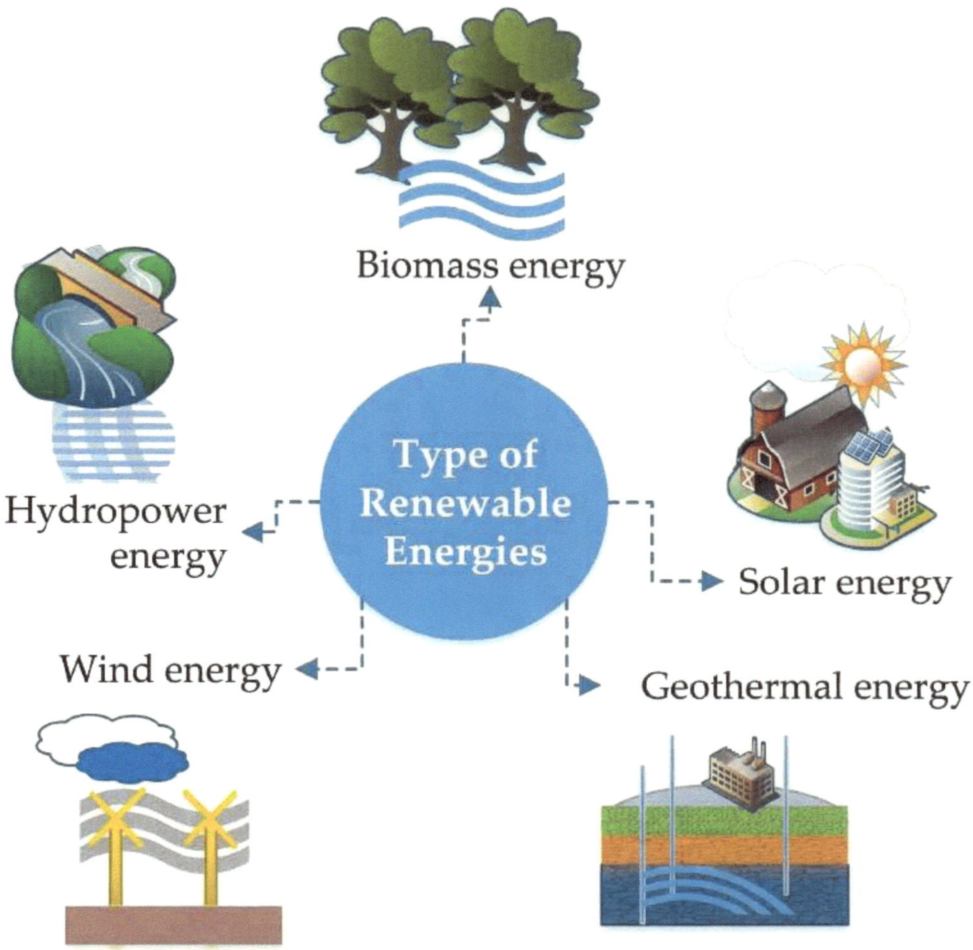


**Fig. (1).** Different types of renewable energy technologies [14].

**Solar Energy:** Solar power uses Photovoltaic (PV) cells or solar thermal systems to collect the sun's radiant energy. It is one of the most abundant and commonly used renewable energy sources. Solar technology has many applications, ranging from modest residential systems to large-scale utility installations.

**Wind Energy:** Wind turbines absorb and transform the kinetic energy of the wind into electricity. Wind power, being one of the fastest-growing energy sources, has enormous potential for large-scale generation, particularly in areas with high wind speeds.

**Geothermal Energy:** Geothermal energy utilizes heat stored beneath the Earth's surface to generate power or give direct heating. This renewable resource is both reliable and capable of generating constant, base-load power, making it highly valuable for energy systems.

**Hydro Energy:** Hydropower uses flowing or falling water to spin turbines, which generate electricity. It is one of the oldest and most well-known sources of renewable energy, and it continues to play an important role in worldwide energy production, particularly in areas with abundant water resources [15].

**Biomass Energy:** Biomass energy is derived from transforming organic matter, including agricultural byproducts, wood, and animal waste, into heat, electricity, or biofuels. This renewable energy source is adaptable and carbon-neutral, providing an eco-friendly solution for waste management while producing energy.

Throughout this chapter, we will explore the fundamentals of each technology, including its underlying mechanisms, advantages, and limitations. In addition, we will examine the current advancements and implementations of these technologies, along with future trends and innovations that could shape their role in a sustainable energy future. Understanding the potential of renewable energy technologies is crucial for meeting the world's growing energy demands and safeguarding the environment for future generations.

## SOLAR ENERGY

### Fundamental Principle

Solar energy involves capturing the sun's radiant energy, which includes visible light, UV rays, and infrared radiation, and converting it into usable power such as electricity or heat [16, 17]. This type of energy can be harnessed and converted into electricity or heat through different technologies [18]. This process relies on harnessing electromagnetic radiation, where photons from the sun interact with materials to transfer their energy. Solar energy systems aim to maximize this energy capture for practical applications like electricity generation and water heating [19, 20].

### Photovoltaic Effect

The Photovoltaic (PV) effect is one of the processes for generating electricity from solar energy that is extensively used. This capability is harvested by using silicon, one of the most utilized semiconductor, materials in the construction of solar cells [21, 22]. In the construction of solar panels, a common design includes two laminated layers of semiconductor materials with different doping types—one n-type (with excess electrons) and the other p-type (with a deficiency of electrons). When sunlight irradiates the surface of the solar cell, the energy from the photons excites the electrons. This energy allows electrons in the n-type layer to become mobile and diffuse across the junction, forming electron-hole pairs (excitons) and enabling the generation of electric current [23]. These free N-type electrons will now flow to search the p polymorphic region, establishing the electrical current direction. This electric current is collected on conductive fabrics on the surface of the cell. Some cells may be connected to active electrical circuits [24, 25].

### Solar Thermal Energy

Another essential principle of solar energy is the conversion of thermal energy, which refers to the capture and use of heat to produce electricity or, in a more direct application, for heating. In the systems built, sun collectors or mirrors are employed to concentrate the sun's rays onto a container with a heat-conductive substance, such as water or oil [26]. This heat can then cause water to boil, creating steam that turns the blades of turbines, which are attached to generators, therefore producing electricity [27]. Another frequent utilization of solar thermal energy is the use of solar water heaters, which employ the sun's energy for heating water [28].

### Energy Transfer and Storage

One of the core problems of solar energy systems is storing or distributing energy for use at times when sunlight is not available, such as during the night or when the sky is overcast. As a rule, electricity generated by photovoltaic systems is stored using batteries, which are mostly lithium-ion batteries [29]. In solar thermal systems for heat storage, there is the option of using molten salts, which can retain solar heat energy and provide energy for long durations, thereby making it possible to provide energy even when the sun is not shining [30, 31].

### Conversion Mechanism

There are two principal methods of transforming solar energy into a useful form of energy, such as Photovoltaic (PV) systems and solar thermal systems [32].

### Photovoltaic (PV) Conversion

Solar cells can generate solar energy in the form of electricity, utilizing solar photovoltaic technology [33]. Once the solar cells are fabricated, they are typically made with this solar technology, which uses silicon and other semiconductor materials. The sunlight contains particles called photons, which free electrons from their atoms when it strikes the solar panel, producing electricity. This process is termed the photovoltaic effect, and it allows energy from solar panels to be generated. Solar installations are remarkably flexible in that they can be built to any diameter, from small units that sit on the roof to huge arrays that cover entire tracts of land. They are also popular for household and business electricity generation because of their high efficiency and declining cost.

### Solar Thermal Energy

The thermal energy of the sun is not transformed into electrical energy directly in this method. The sun's rays are reflected and concentrated using mirrors, lenses, or collectors in solar thermal systems, which in turn produce thermal energy to heat a heating medium or heated liquid, such as water or molten salts [34]. This energy can also be utilized for heating water or spaces. In a more sophisticated application, the remarkable solar concentrator power systems use thermal energy to create steam, which turns a turbine to create electricity. Regions of high direct solar heat energy allow for effective CSP systems, and their thermal storage feature enables energy production even at nighttime [35].

The development of solar energy technologies requires these conversion mechanisms, which offer different ways to harness renewable energy depending on the purpose and location.

### Advantages and Limitations

#### *Advantages of Solar Energy*

- **Renewable and Abundant:** Solar energy is a universal energy solution, providing power as long as the sun shines, even in regions with lower sunlight levels.
- **Environmentally Friendly:** Solar power is a clean energy source that significantly reduces carbon footprint compared to fossil fuels, producing no direct greenhouse gas emissions, air pollution, or water contamination.
- **Reduced Energy Bills:** Solar panels can significantly reduce electricity bills by directly offsetting grid power consumption, and in some cases, surplus energy can be reinvested into the grid.
- **Energy Independence:** By producing their own electricity, individuals, companies, and countries can reduce their reliance on non-renewable energy sources and mitigate vulnerability to fluctuating energy costs.
- **Low Maintenance Costs:** Solar energy systems require minimal maintenance, with durable panels and long warranties. Regular cleaning and occasional checks on the inverter are sufficient for optimal performance.

- **Diverse Application:** Solar energy serves multiple purposes, including electricity generation through photovoltaic systems, heating water, and powering devices in remote areas without traditional grid access.

*Limitations of Solar Energy*

- **Intermittency:** Solar energy, which relies on sunlight, is unavailable at night or during cloudy weather, necessitating storage solutions or supplementary energy sources to ensure continuous power production.
- **High Initial Investment:** Despite a decrease in initial costs for solar panels and solar thermal systems, large-scale installations still require substantial upfront expenses.
- **Energy Storage Challenges:** Solar energy production during peak electricity demand may not align, necessitating sophisticated energy storage systems for future use, and potentially increasing the total cost.
- **Space Requirements:** Large-scale solar installations, particularly solar farms, require extensive land for energy production, making it challenging to find suitable space in densely populated or urban areas.
- **Weather Dependency:** Local weather conditions significantly impact solar energy production, with areas with frequent cloud cover or limited sunlight potentially unable to produce solar power consistently or in large quantities.
- **Energy Conversion Efficiency:** Solar panels' efficiency has improved, but their energy conversion rate remains lower than other energy sources, as typical photovoltaic panels only convert a portion of the sunlight into electricity.

**Current Advancements and Implementations**

In recent years, solar energy technology has progressed significantly, with innovations enhancing both its efficiency and availability. Below are some of the key developments and their real-world implementations:

**Enhanced Photovoltaic (PV) Efficiency**

Research in solar cell technology has increased efficiency rates, allowing more sunlight to be converted into electricity. There are new materials, which include perovskite solar cells, that are viewed as promising materials in the manufacturing of cells that are silicon-based. These materials are cheaper and more efficient, which in turn promotes the use of solar power. Nowadays, more and more people are using bifacial solar panels, which can catch sunlight from both sides, generating more energy using the same space.

### Cost Reductions

The price of solar power has greatly reduced over the last 10 years, which has put it in direct competition with fossil fuels [36]. This is especially true in the case of utility-scale solar projects, where large-scale installations are among the cheapest sources of power today. Other contributing factors include reductions in the price of manufacture, the availability of improved supply chains, and economies of scale, which increase solar energy accessibility to both residential and commercial users.

### Energy Storage Solutions

As the problem of intermittency still stands as one of the barriers, progress in technology related to energy storage units has been shown to be significant. To store solar energy, which is essential so that energy can be consumed even when the sun is not shining, lithium-ion batteries are often used. For this reason, energy outages can also be avoided, thus achieving a stable supply of electricity. Advances in the development of battery systems and domestic units for energy storage allow the use of solar energy in everyday energy consumption, as it can be utilized at any time of day.

### Concentrated Solar Power (CSP)

A newer technology that utilizes mirrors or lenses to focus sunlight and create thermal energy is called CSP. Energy storage can also be enhanced by considering that modern CSP plants use molten salt as a heat transfer fluid, enabling electricity generation even when the sun goes down. Thus, this advancement increases the dependability and expandability of solar energy use in such regions where large-scale solar energy systems are built.

### Floating Solar Farms

The application of floating solar farms is one such technique where solar energy can be generated by deploying panels on water bodies rather than land. These systems can be deployed on water bodies, including water reservoirs, lakes, or oceans, which reduces the space required for conventional solar power grids. Another advantage of floating solar farms is that they provide temperature control, which improves efficiency and reduces water evaporation. China, Japan, and India are among other countries that have set forth these systems to minimize land usage.

These developments indicate that solar energy is on its way from being a niche application to a ubiquitous one that finally joins the race for the world's sustainable energy. With much more development taking place in the solar industry, it is set to become an even more important component of energy systems in the present as well as the future.

## Future Trends and Innovations

Solar energy is on an optimistic path, marked by further technological development and a growing need for clean energy worldwide. Below are some of the major trends and innovations anticipated to influence the future development of solar power:

### Increased Efficiency through Advanced Materials

The quest for new materials, such as perovskite solar cells, is one of the most interesting areas of research in the solar technology space. Such materials are expected to perform better than or be more cost-effective and easier to manufacture than silicon-based panels. The combination of perovskite with silicon, which is called a tandem solar cell, can enable it to surpass the efficiency of panels greater than 30 percent; thus, solar energy can be more attractive than other sources of energy.

### Flexible and Lightweight Solar Panels

Thin-film and flexible solar panels are expected to increase the functional characteristics of future solar panels and allow them to be connected to more components. These panels are lightweight in nature and can be used on surfaces where traditional panels would never be used – such as on vehicles and portable devices. The prospect of integrating flexible panels into everyday products, such as clothing and backpacks, has created new frontiers for solar energy that is portable and renewable.

### Energy Storage and Smart Grid Integration

One of the key factors that has impeded the use of solar energy harnessing has been its lack of availability. However, advances in energy storage, particularly in batteries, are making solar power more dependable. Storage solutions for grids, such as large-sized lithium-ion batteries, in addition to newer technologies, such as solid-state batteries and flow batteries, will allow us to store surplus solar energy and use it when it is dark. Furthermore, integration with smart grids will help distribute solar energy in a more efficient manner and allow excess energy from homes and businesses to be fed back into the grid for better demand management.

### Artificial Intelligence (AI) and Automation in Solar Systems

Adding AI and automation to solar energy systems will improve the amount of energy produced and the level of efficiency in the management of the entire system. AI can predict future weather conditions, assist in the proper tilt of solar panels to increase exposure to the sun, and control energy storage as well as the distribution. Drones, for example, can be used for the maintenance of large installations such as solar farms, thus contributing to improved operational efficiency and cost savings.

### Hybrid Renewable Energy Systems

The emerging trend in the industry is not to rely on a single energy source, but to combine solar energy with other renewable energy sources, such as wind, hydro, or biomass, to form hybrid systems that are more reliable than single sources because one source can offset the variability of others. A solar-wind hybrid system would make use of both sunny and windy days to meet energy requirements without using fossil fuels as much.

These innovations are driving the solar industry toward greater efficiency, affordability, and widespread adoption, positioning solar energy as a cornerstone of the future global energy landscape.

Having explored solar energy, a technology highly dependent on sunlight availability, we now turn to wind energy, which can generate power under different atmospheric conditions and complement energy's limitations.

## WIND ENERGY

### Fundamental Principle

Wind energy captures the kinetic energy of air in motion and transforms it into mechanical or electrical power that can be utilized. The fundamental principle behind wind energy is based on the conversion of this kinetic energy into mechanical energy using wind turbines. When the wind flows, it drives the rotation of a wind turbine's blades. These blades are connected to a rotor, which in turn is linked to a shaft. As the blades spin, the rotor turns the shaft, converting the wind's kinetic energy into mechanical energy, as shown in Fig. (**2**). In contemporary wind turbines, the mechanical energy is transferred to a generator, transforming it into electrical energy for powering homes, businesses, and other facilities. The amount of energy that can be captured from wind depends on several factors, including wind speed, air density, and the swept area of the turbine blades. According to Betz's law, no wind turbine can capture more than

59.3% of the kinetic energy in wind, which is known as the Betz limit, due to energy losses in the conversion process. The energy available in the wind increases dramatically with wind speed, as it is proportional to the wind speed raised to the third power. This means that even slight gains in wind speed can lead to substantial boosts in energy production. Because of their favorable characteristics, regions with consistent, forceful winds, such as coastal locations and vast plains, are frequently chosen as sites for wind farms.

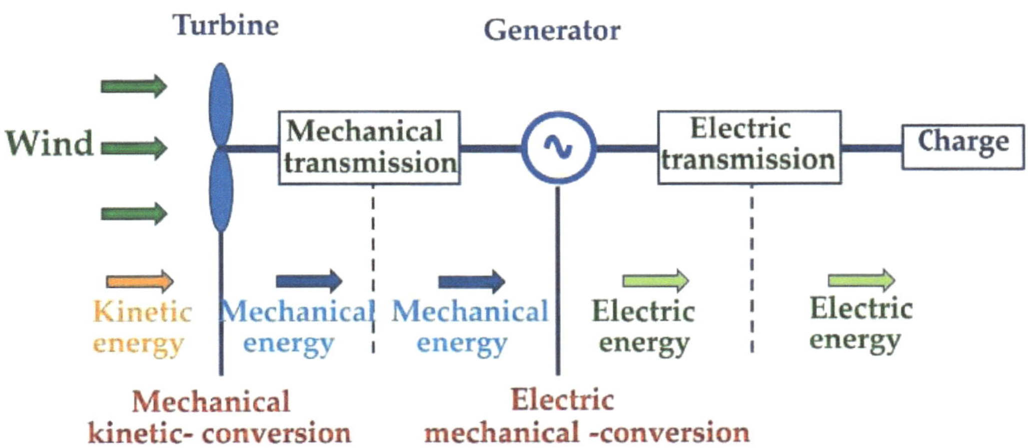


**Fig. (2).** Fundamental principle of wind energy conversion [37].

## Conversion Mechanism

Wind energy is primarily converted into usable power through wind turbines, which capture the kinetic energy from the movement of air and transform it into electrical energy. The fundamental idea behind this process is to capture the energy from wind movement and convert it into mechanical energy, which is then transformed into electrical energy. The key conversion mechanisms in wind energy involve the following steps:

## Wind Capture by Turbine Blades

The process begins when wind flows across the turbine blades, which are engineered to harness the kinetic energy from the air in motion. The blades function similarly to an airplane wing, utilizing an aerodynamic design that creates a pressure difference on either side of the blade as wind passes over it [38]. This pressure difference generates lift, causing the blades to rotate. Modern wind turbines typically have three blades mounted on a rotor, and their size and shape are optimized to capture as much wind energy as possible. Other factors that influence the amount of energy harvested include the wind velocity, air density and rotor swept area which is defined as the area enclosed by the rotating blades [39]. The larger the swept area; and the higher the wind velocity, the more the energy directed to the system.

### Mechanical Energy Generation

The rotor situated at the centre is responsible for the spinning motion which in turn causes the connected shaft to rotate. In essence, the kinetic energy present in wind is converted to mechanical energy during this stage. This mechanical energy, which most of the time is incorporated in a gearbox system, emanates from the rotor; and is directed towards a generator. There are some turbine designs that do not contain a gearbox, known as direct-drive turbines, where the shaft is directly linked to the generator. Most of the time, however, a gearbox increases the RPMs to allow the generator to reach more effective operating speeds in large-sized wind turbines [40].

### Power Conditioning and Grid Integration

When electricity is generated, it is fed to the power grid, but this only happens after conditioning has been done on it. The term 'power conditioning' refers to various operations, such as voltage regulation and frequency adjustment, to make sure that the generated electricity is in line with grid standards. This is a very important measure that enables wind energy to be used in homes, industries, businesses, and other structures, which helps in preserving the grid. For wind plants that are commercially sourced, the electricity produced is delivered over long distances to the users. This requires high-voltage transformers and transmission lines to minimize energy loss during distribution.

### Energy Storage

Due to the intermittent nature of wind energy, as it relies on the wind's availability and strength, some systems incorporate energy storage solutions to maintain a consistent power supply. In these configurations, surplus electricity produced during strong wind conditions is stored in batteries or other energy storage systems, allowing it to be used later when wind speeds are insufficient [41]. While energy storage is not a direct part of the conversion mechanism, it plays a critical role in balancing supply and demand in wind energy systems [42].

### Advantages and limitations

#### *Advantages of Wind Energy*

- **Renewable and Abundant:** Wind energy, a renewable resource, generates electricity if wind currents persist, playing a crucial role in various global regions.

- **Environmentally Friendly:** Wind energy is a clean and environmentally friendly alternative to fossil fuels, reducing reliance on these sources and thereby contributing to climate change efforts [43, 44].
- **Low Operating Costs:** Wind turbines offer a cost-efficient energy solution over time due to their minimal fuel usage and lower ongoing operation and maintenance costs compared to conventional power plants.
- **Scalability:** Wind energy projects, ranging from individual turbines to large farms, offer versatility in various settings, from small rural areas to large utility-scale developments.
- **Energy Independence:** Wind energy enhances energy security by broadening the energy mix and lowering reliance on imported fossil fuels, thereby promoting greater energy independence in nations with significant wind resources.

*Limitations of Wind Energy*

- **Intermittency:** Wind energy faces challenges due to erratic wind patterns, requiring backup systems or energy storage solutions to maintain a reliable power supply [45].
- **Visual and Noise Impact:** Wind turbines significantly impact landscape aesthetics, particularly in rural areas, and cause disruptive noise, although modern designs have significantly reduced noise levels [46].
- **Wildlife Impact:** Research is exploring ways to mitigate wind turbine threats to birds and bats, such as modifying turbine design or locating wind farms away from critical wildlife habitats.
- **High Initial Costs:** Wind turbine construction and installation can be costly, requiring significant upfront investment in infrastructure, transmission lines, and grid connection, despite low operational costs.
- **Space Requirements:** Large-scale wind farms require significant land or offshore areas for efficient operation, potentially leading to disputes over land use, particularly in areas with limited or high agricultural, recreational, or scenic values [47].

## Current Advancements and Implementations

Wind energy, which is recognised as one of the fastest-growing forms of renewable energy, is constantly evolving, with significant advances in technology, design, and efficiency. As countries and industries seek cleaner energy alternatives to reduce carbon emissions and dependence on fossil fuels, wind power is playing a crucial role in the global energy transition [48]. Recent developments are enhancing both onshore and offshore wind capabilities, making wind energy more competitive and scalable. Below are some of the key advancements and implementations in wind energy:

## Larger and More Efficient Turbines

A key trend in the wind energy sector is the advancement of larger and more efficient turbines. Modern wind turbines are now capable of generating more electricity due to their increased size and improved blade designs. For example, newer models such as the Haliade-X by GE and Vestas V236-15.0 MW feature rotor diameters exceeding 200 meters and can produce more than 12–15 MW of power per turbine. These larger turbines are particularly effective in offshore installations, where wind speeds are higher and steadier.

- **Longer Blades:** Blade design improvements, including carbon fiber and aerodynamics, have led to longer, wind-efficient blades, reducing energy output and lowering Kilowatt-hour (kWh) costs.
- **Direct Drive Systems:** Direct-drive turbines offer increased reliability, reduced maintenance costs, and increased durability, especially in harsh environments, by eliminating the need for gears.

## Floating Offshore Wind Farms

Traditionally, offshore wind farms have been established in shallow waters utilizing fixed-bottom structures. However, the emergence of floating offshore wind turbines has enabled the capture of wind energy in deeper waters, where wind speeds tend to be higher and more consistent. Floating wind technology involves turbines mounted on platforms that are anchored to the seabed using mooring lines, allowing them to operate in depths where fixed foundations are impractical.

- Hywind Scotland and WindFloat Atlantic are pioneering floating offshore wind farms, demonstrating the feasibility of cost-effective, scalable floating platforms to unlock vast wind resources in deep coastal regions.

## Advanced Energy Storage Integration

One of the obstacles associated with wind energy is its variability; the wind does not consistently blow when energy demand peaks. To address this, energy storage systems, especially battery technologies, are being integrated into wind farms to store excess energy and provide power during low-wind periods. This improves the reliability of wind energy and ensures a steady power supply to the grid.

- **Hybrid Wind-Storage Systems:** Wind farms are integrating battery storage into hybrid systems, allowing power generation flexibility; and reducing fossil fuel dependency. Cost-effective solutions, like Hornsdale Power Reserve in Australia, are expected to follow suit.

### Digitalization and Smart Wind Farms

The development of digital technologies, such as AI, machine learning, and the IoT, is showing innovations in the wind energy sector. With these developments, wind farms can operate more efficiently and have predictive maintenance and real-time management.

- **AI and Machine Learning:** Wind turbine operators deploy AI based algorithms to both anticipate the wind directions and align the turbine. AI also contributes to the accuracy of the estimation of maintenance costs [49].
- **IoT Sensors and Drones:** IoT sensors are now being used to track turbine health parameters in real-time including pitch, wind speed, and vibration. Drones are also being adopted for the purpose of inspections allowing operators to visually see the turbine and its thermal components.

### Repowering and Life Extension of Existing Wind Farms

As wind turbines age, many older wind farms are being repowered or upgraded to extend their operational life and improve efficiency. Repowering refers to the process of substituting outdated turbines with advanced, more efficient versions or enhancing components, like blades or control systems, to improve overall performance.

- **Repowering Projects:** In the United States and Europe, many wind farms are undergoing repowering projects [50]. For example, replacing a 1 MW turbine installed in the 1990s with a modern 3 MW turbine can triple the energy output from the same site. Repowering is seen as a cost-effective way to increase renewable energy capacity without needing additional land or environmental permits.

### Offshore Wind Expansion

Offshore wind is experiencing rapid growth, particularly in Europe and Asia, where governments are investing heavily in large-scale wind farms. Nations such as the United Kingdom, China, and Germany are at the forefront of offshore wind capacity development. These projects not only harness strong ocean winds but also contribute to coastal economies by creating jobs in manufacturing, installation, and maintenance.

- **Floating Wind Innovations:** Innovations in floating wind technology enable offshore farms to be deployed farther from shore, increasing offshore wind energy potential in previously inaccessible areas like the Mediterranean, California, and Portugal.

### Small-Scale Wind Turbines and Distributed Energy

While utility-scale wind farms dominate the sector, small-scale wind turbines are becoming more popular for residential, agricultural, and industrial use. These turbines, generally available in capacities from 1 kW to 100 kW, offer a decentralized energy option, enabling both individuals and businesses to produce their own electricity.

- **Distributed Wind Energy:** Distributed energy resources (DERs), including small-scale wind, solar, and storage systems, are revolutionizing the energy landscape by providing reliable, sustainable power in rural areas and isolated communities.

### Future Trends and Innovations

As wind energy technology advances, numerous emerging trends and innovations are poised to influence its position within the global energy framework. These advancements aim to improve efficiency, reduce costs, and expand the potential for wind energy generation in various environments. Key trends and innovations include:

### Hybrid Wind and Solar Systems

Hybrid renewable energy systems that combine wind and solar power are emerging to enhance energy production and stability [51]. These systems leverage the synergistic benefits of both wind and solar energy generation, as wind often blows when the sun is not shining, particularly at night or in cloudy conditions. By integrating wind turbines with solar panels, energy providers can create a more reliable power supply that reduces intermittency. Hybrid systems can be implemented in both large-scale power facilities and decentralized, off-grid applications, providing clean energy in remote or underserved areas.

### Advanced Energy Storage Solutions

The variability of wind energy has historically posed significant challenges, but recent advancements in energy storage technologies are making strides in mitigating this problem. The development of next-generation batteries, such as solid-state and flow batteries, is allowing wind farms to store excess energy generated during periods of high wind activity for use during times when wind is less abundant. Green hydrogen production is, on the other hand, becoming widely accepted as an option for harnessing wind energy for end-use apart from battery storage. This is the case as the hydrogen produced through electrolysis systems can be directly utilized whenever there is a surplus of wind energy. This hydrogen can be stored and applied at a future time for the generation of electric power; to fuel vehicles or to support the industry, thus providing a reliable and green solution for energy storage.

### Artificial Intelligence and Predictive Analytics

Wind energy O&M (Operation and Maintenance) processes are undergoing a fundamental change due to the incorporation of AI and machine learning technology. AI algorithms can analyze and utilize numerous practical data sets, including wind turbine rotational analysis, climate and wind data, and demand for electric supply in the grids, to optimize energy output while predicting faults on wind turbines before they happen. Given its qualitative advantages, such predictive maintenance enhances efficiency by minimizing downtime, which in turn increases the working lifetime of wind turbines. The use of predictive maintenance tools enables operators to identify when wear or breakdown occurs and perform corrective action before the entire system becomes defective. Economically speaking, this helps cut down the number of repairs required and optimizes the supply chain.

### Wind Turbine Recycling and Sustainability

Over the years, wind turbine technology has advanced quite a bit. The disposal of wind turbines has, however, gained attention as an important issue. Parts of wind turbine blades, which are created using composite materials, are challenging to recover through recycling, and in some instances, have been disposed of in landfills. To resolve this issue, there is a research and industrial effort meant to source blade materials and processes that can be easily reused. It is also reported that blade designs using thermoplastic composites are currently being developed that can easily be remelted. Besides, circular economy strategies are being developed where some parts of dismantled wind turbines are reused in the construction of other wind turbines.

### Smart Grid Integration

One of the major highlights of wind energy in the future is that it will go along with the development of smart technology. The supply and demand of electricity can be controlled in real time thanks to the integration of advanced communication technology in the grids. There is an advancement in solar and wind energy; to enhance the integration of smart grid technologies with renewable energy for better performance. By using smart meters and real-time data, grid operators can adjust power flows to match variable wind energy production, reducing the risk of blackouts or energy shortages. This innovation will make wind energy more reliable and responsive to changing grid conditions.

While wind energy is shaped by air movement patterns, geothermal energy provides a more stable and constant energy source derived from the Earth's internal heat.

## GEOTHERMAL ENERGY

### Fundamental Principle

Geothermal energy is sourced from the heat contained beneath the Earth's surface, which is a result of the planet's initial formation and the continuous radioactive decay of elements found in the mantle and crust. The essential concept behind geothermal energy is to utilize this internal heat for electricity generation or direct heating applications [52]. Essentially, geothermal energy capitalizes on the natural temperature differences within the Earth. The Earth's internal heat is conducted towards the surface, creating regions where heat is more accessible. This geothermal heat can be captured and utilized through various methods:

### Geothermal Heat Pumps

These systems leverage the consistent temperature of the shallow ground to offer heating during winter and cooling in summer [53]. A geothermal heat pump circulates fluid through underground piping, as illustrated in Fig. (**3**), to either absorb heat or release it, depending on the season. This method leverages the Earth's relatively constant subterranean temperature, which remains more moderate compared to surface temperatures.

### Geothermal Power Plants

These facilities produce electricity by utilizing steam generated from heated underground water. The steam drives turbines connected to generators. There are several types of geothermal power plants:

- **Dry Steam Plants:** These directly utilize steam from geothermal reservoirs to turn turbines.
- **Flash Steam Plants:** These take on high-pressure hot water from geothermal reservoirs, allow it to flash into steam, and use this steam to drive turbines.

- **Binary Cycle Plants:** These systems utilize heat from geothermal hot water, transferring it to a secondary fluid with a lower boiling point. This secondary fluid vaporizes as a result, and the resulting vapor is used to drive turbines.

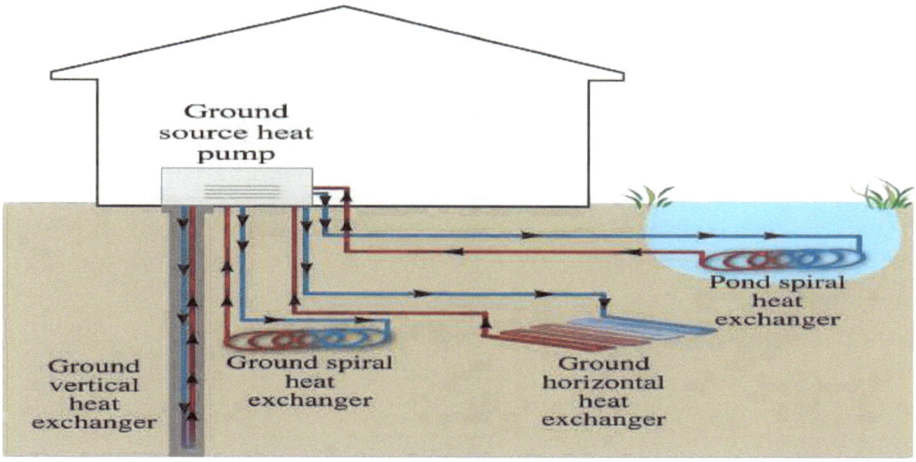


**Fig. (3).** Schematic diagram of geothermal heat pumps [54].

## Direct Use Applications

This involves using geothermal hot water directly for applications such as district heating, agricultural processes, and industrial uses. Hot spring water or geothermal fluids are extracted and used without the need for a heat conversion process. The effectiveness of geothermal energy depends on accessing geothermal resources with sufficient heat and fluid flow. Common geothermal resources include geothermal reservoirs, hot springs, and volcanic regions where heat is more readily available. By exploiting these natural heat sources, geothermal energy provides a sustainable and reliable source of power with minimal environmental impact.

## Conversion Mechanism

Geothermal energy has been harvested for a long time, mostly in the form of direct energy for heating spaces; today, its application is more diversified and includes producing energy. Geothermal energy can be characterized as environmentally friendly energy, which is renewable and has great possibilities as it is obtained from the heat of the Earth. The use of geothermal energy usually means the extraction of heat located deep in the Earth's crust for use as electricity, warmth for buildings, or for a variety of other industrial applications.

## Direct Use of Geothermal Heat

In areas where geothermal heat sources are abundant, the most practical solution is to have direct heat intervention for different purposes. This technique involves tapping geological heat reservoirs with hot water or steam and utilizing the same directly without the need for an intermediary conversion. Direct applications of this method can be observed in various contexts, including the following:

- **Geothermal Heating:** Hot water from geothermal sources can be piped into buildings for space heating. This trend is especially pronounced in places like Iceland, where the availability of geothermal resources allows for easy implementation.
- **Industrial Applications:** Geothermal heat is used in industrial activities that demand a lot of heat, such as drying agricultural products, pasteurization, and greenhouse heating.
- **Aquaculture and Spa Heating:** Geothermal energy is incorporated into aquaculture centres and is used in recreation at hot spring resorts and spas.

The advantage of the direct use method is that it does not involve high energy waste during the conversion processes; however, one of its limitations is that it can only be used in areas with rich geothermal resources.

## Electricity Generation Using Geothermal Energy

In general, the main process of the conversion of geothermal energy into electricity involves transporting heat in the geothermal reservoir and sequentially transforming a part of the thermal energy into electric energy as a result of thermodynamic processes. Geothermal power plants mainly use three types of technologies designed to work in given temperatures and resource conditions.

- **Dry Steam Power Plants:** The dry steam plant stands out as the most basic and conventional method of generating geothermal electricity. Steam originating from geothermal reservoirs is used to spin a turbine attached to a generator. The wells extract steam, which is then transported to the power plants where it is used to produce electricity. After its use, the steam is cooled into water and pumped back into the ground to preserve the pressure of the reservoir. This practice is common in areas endowed with high temperature geothermal resources like the Geysers in California.
- **Flash Steam Power Plants:** Flash steam plants constitute the major proportion of Geothermal Power facilities. These systems use high-pressure hot water from geothermal wells, which is passed into a low-pressure chamber. The drop in pressure causes a portion of the water to "flash" into steam. This steam is then

used to drive a turbine, generating electricity. The remaining liquid water is either flashed again or reinjected into the reservoir. Flash steam plants are particularly effective in areas with high-temperature geothermal resources above 180°C.

- **Binary Cycle Power Plants:** Binary cycle plants are used in lower-temperature geothermal fields where the geothermal fluid is not hot enough to flash into steam on its own [55]. Instead, the geothermal water is used to heat a secondary working fluid with a lower boiling point, such as an organic compound or a refrigerant. The secondary fluid vaporizes and drives a turbine to generate electricity. This closed-loop system allows geothermal plants to operate in areas with moderate temperatures, expanding the potential use of geothermal resources. Binary plants are highly efficient and environmentally friendly, as there is minimal release of geothermal fluids into the atmosphere.

## Enhanced Geothermal Systems (EGS)

EGS represents an advanced approach to geothermal energy conversion. EGS technologies aim to expand the use of geothermal energy by creating artificial reservoirs in areas where natural geothermal resources are insufficient or absent. This is done by drilling deep into the Earth's crust, fracturing the rock, and injecting water to create or enhance geothermal reservoirs, as shown in Fig. (**4**). The injected water is heated by the Earth's natural heat, and the resulting hot water or steam is extracted and used to generate electricity, much like in conventional geothermal power plants [56]. EGS holds the promise of significantly expanding the geographic availability of geothermal energy, potentially making it a widespread renewable energy source. Although EGS is still in the experimental and pilot stages, it represents a major opportunity for future geothermal energy development.

## Geothermal Heat Pumps (Ground Source Heat Pumps)

Geothermal heat pumps, or ground source heat pumps, use the consistent temperature found just a few meters beneath the Earth's surface to supply heating and cooling to buildings. Unlike other geothermal systems that rely on deep, high-temperature reservoirs, heat pumps use the constant temperature of the shallow subsurface, which remains relatively stable throughout the year. During the winter, geothermal heat pumps extract heat from the ground and transfer it into buildings [58]. During the summer, the system works in reverse, extracting heat from the building and releasing it into the cooler ground. This highly efficient system is becoming increasingly popular for residential, commercial, and industrial heating and cooling applications. Geothermal heat pumps are highly energy-efficient and cost-effective over time because they require only a minimal amount of electricity to function. However, the upfront installation costs can be significant, primarily due to the need for drilling and setting up underground loops [59].

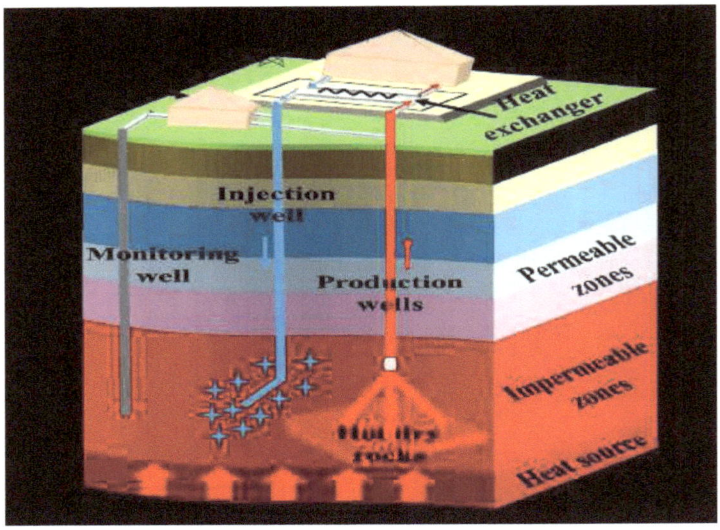


**Fig. (4).** Schematic diagram of enhanced geothermal systems [57].

## Advantages and limitations

### *Advantages of Geothermal Energy*

- **Renewable and Sustainable:** Geothermal energy comes from the heat generated within the Earth's core, a source that is constantly created and renewed. As long as the Earth exists, this energy source will remain available, making it highly sustainable and renewable over long periods.
- **Low Greenhouse Gas Emissions:** Compared to fossil fuels, geothermal energy has a very low carbon footprint. While there may be minor emissions associated with the extraction process, they are significantly lower than those from traditional energy sources like coal or natural gas.
- **Small Land Footprint:** Geothermal power facilities require significantly less land than solar or wind farms. This makes them acceptable for regions with limited available space, and they frequently have a smaller impact on nearby ecosystems.
- **Energy Independence:** Nations with large geothermal resources can lessen their reliance on imported fossil fuels, thereby improving energy security. This energy autonomy minimizes the volatility of the global energy markets and fuel prices.
- **Economic Benefits:** Geothermal energy projects contribute to local economic growth by generating employment opportunities in areas such as construction,

operation, and maintenance. Additionally, the stable energy supply can help stabilize electricity prices in regions that adopt geothermal power generation.

### *Limitations of Geothermal Energy*

- **Location-Specific:** One significant drawback of geothermal energy is its reliance on specific geographic locations. It is most viable in regions with significant tectonic activity, such as the Pacific "Ring of Fire." Countries or areas without easy access to geothermal reservoirs may find it challenging or impossible to implement geothermal energy systems.
- **High Initial Costs:** The upfront costs for geothermal projects can be quite high due to the need for exploration, drilling, and specialized infrastructure. Locating suitable geothermal reservoirs requires extensive geological surveys, and drilling deep wells can be both time-consuming and expensive. However, once a plant is operational, the long-term costs are generally low.
- **Environmental Concerns:** Although geothermal energy is largely viewed as environmentally friendly, there are some issues to consider. Extracting geothermal fluids can lead to the release of trace amounts of gases such as hydrogen sulfide and carbon dioxide. In addition, if not properly managed, the process can lead to water contamination, land subsidence, or seismic activity (induced earthquakes), particularly in areas where the Earth's crust is more fragile.
- **Resource Depletion Risks:** While geothermal energy is renewable, there is a risk of local resource depletion if the geothermal reservoirs are not properly managed. Over-extraction of geothermal fluids without adequate replenishment can lead to a reduction in the temperature or pressure of the geothermal resource, reducing the efficiency of energy production over time.
- **Energy Transmission Challenges:** In regions where geothermal energy is available but far from population centres, the cost of transmitting electricity over long distances can be high. This challenge can limit the feasibility of geothermal energy in more remote areas unless substantial infrastructure investments are made.

## Current Advancements and Implementations

Geothermal energy, derived from the Earth's internal heat, is gaining prominence as a reliable and sustainable energy source. Recent advancements in geothermal technology have expanded its scope and improved efficiency, while new implementations are helping it play a more significant role in the global energy mix.

## Enhanced Geothermal Systems

A breakthrough in geothermal energy is the creation of Enhanced Geothermal Systems (EGS). Traditional geothermal systems require naturally occurring hydrothermal resources, including heat, water, and permeable rock formations. EGS, however, overcomes this limitation by creating engineered reservoirs where natural conditions are insufficient. EGS refers to a process where water is injected into hot, dry rock formations where it absorbs heat and is brought to the surface to produce electricity. This technology compensates for the geographic limitation of producing geothermal energy in volcanic or tectonically active regions. Countries that did not have the capacity to consume geothermal energy are now expanding and undertaking EGS technological advancement.

## Binary Cycle Power Plants

The emergence of binary cycle power plants has systematized the utilization of geothermal energy. Geothermal energy can now be harnessed from relatively lower temperature resources to produce electrical energy [60]. In these systems, the geothermal water is brought into contact with a secondary fluid with a lower boiling point, which vaporizes to drive a turbine and produce electricity. This approach increases efficiency, ensures that new geothermal reservoirs that were not previously reachable are more accessible, and is much better for the environment in that it has a closed-loop system to contain emissions and fluids [61]. This development has led to the establishment of new geothermal plants in regions with moderate temperatures, thereby expanding the applications of geothermal energy.

## Direct Use Applications

Distinct from geothermal electricity, geothermal energy is now being integrated into direct-use applications such as agriculture, aquaculture, and community heating systems. Countries like Iceland and New Zealand use geothermal heat for community heating systems, while advancements in technology make these systems viable in colder areas. Geothermal is also used in greenhouses for farming and in aquaculture for temperature control.

## Implementation in New Regions

Geothermal energy was once ipso facto limited to countries in the eastern part of Africa, such as Kenya and Ethiopia, which have now established themselves as key players in geothermal energy production. Eastern Africa was not believed to have much potential for geothermal energy production. Other regions in Europe and Asia are also beginning to include geothermal energy in their renewable energy strategies. In Europe, it is worth noting that Turkey has rapidly advanced its geothermal energy production, somewhat propelled by supportive government regulations and advanced exploration methods.

### The Role of Government Policies and Incentives

The deployment of geothermal energy could have been hampered without government measures like tax exemptions, feed-in tariffs, and finance grants aimed at fostering the development of the sector. Many countries are now adopting policy measures that aim at increasing the share of geothermal energy as part of the renewable energy mix.

### Future Trends and Innovations

Geothermal energy has been around for a long time, like a mature renewable resource, but the technology and exploration continue to improve. Such advancements strive to augment the efficiency, availability, and total input of geothermal energy into the world's energy supply. Global geopolitical changes and many inventions are setting directions for the development of geothermal energy in the future.

### Hybrid Renewable Energy Systems

An emerging trend is the combined use of geothermal energy with solar and wind, which increases the stability of energy systems. In hybrid systems, variable output from the wind and sun is paired with the dependable baseline energy from geothermal, thus a balanced energy supply model is maintained.

### Small-Scale and Modular Geothermal Plants

Due to a rise in demand for decentralized energy solutions, small and modular geothermal power plants are now becoming more popular. These systems allow off-grid or remote areas to receive consistent renewable energy for their power requirements and can be expanded as energy requirements change. Because of the ease of their installation and minimum maintenance requirements, they allow for rapid deployment and reduce adverse effects on the environment thus making geothermal energy more versatile and easier to use [62].

### Geothermal Desalination

Innovative uses of geothermal energy include its application in the desalination process, such as in plants that remove salt from seawater to create fresh water. This is quite useful, especially in areas that are dry and water is scarce. Geothermal desalination addresses the issue of available energy from renewable resources and the need for freshwater simultaneously, providing a two-pronged solution to pressing world problems.

### Sustainable Geothermal Systems

The future development of geothermal energy will be more environment-friendly. The use of closed-loop systems that do not draw on freshwater resources reduces contamination risk and pollution while also minimizing water consumption. Measures to control emissions, for instance, hydrogen sulphide, also make sure that geothermal energy remains one of the cleaner sources out there.

### Global Expansion of Geothermal Projects

Geothermal energy expands beyond traditional countries such as Iceland, the USA, and the Philippines into new regions such as Africa, South America, and Southeast Asia as more countries invest in it. Governments and international organizations are offering financial incentives, research support, and partnerships to encourage geothermal development. This worldwide growth will be crucial in addressing the increasing energy demands globally while decreasing dependence on fossil fuels.

Moving from subsurface heat sources, we now explore hydro energy, one of the most established methods for generating renewable electricity through water flow.

## HYDRO ENERGY

### Fundamental Principle

Hydro energy, or hydroelectric power, is derived from the movement of water, and it represents one of the ancient and most established forms of renewable energy [63]. The fundamental principle behind hydro energy lies in harnessing the kinetic and potential energy of flowing or falling water to produce electricity as shown in Fig. (**5**). At its core, hydro energy operates based on the following principles:

### Energy Source

Water, or the natural movement of water in rivers, streams, or waterfalls, is the primary source of hydro energy. This movement is driven by the gravitational force of the Earth, which causes water to flow from higher elevations to lower elevations.

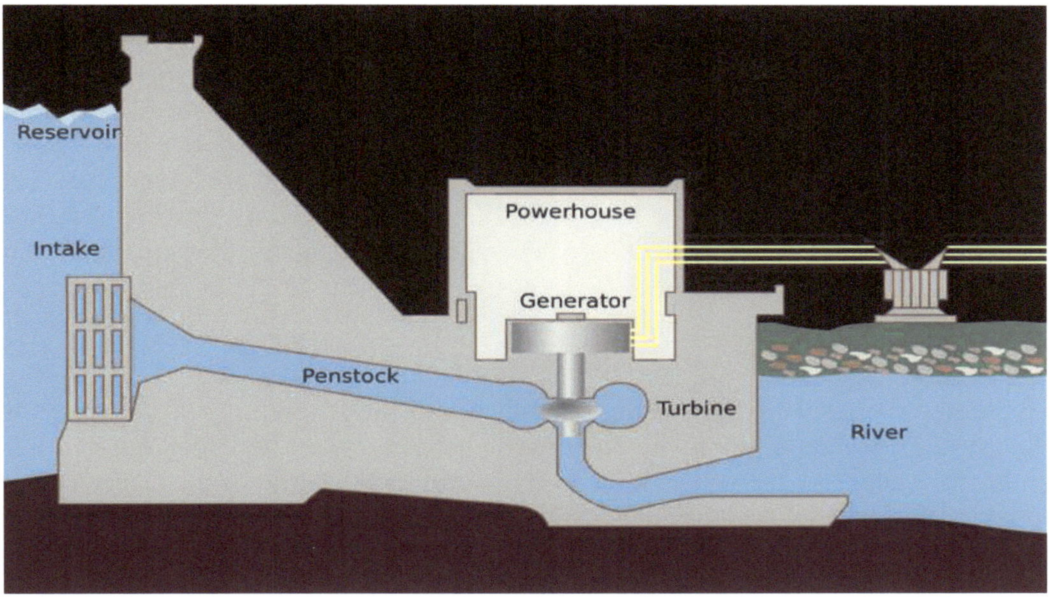


**Fig. (5).** Schematic diagram of hydro energy system [64].

## Potential Energy

When water is stored in a reservoir behind a dam, it gains potential energy due to its elevated position. This potential energy is a result of gravitational forces acting on the mass of the water.

## Conversion Mechanism

To transform potential energy into mechanical energy, water is discharged from the reservoir and permitted to flow through turbines. The movement of the water exerts a force that spins the turbine blades, converting the kinetic energy of the flowing water into mechanical energy.

## Electricity Generation

The spinning turbines are connected to a generator, where the mechanical energy is converted into electrical energy. The generator produces electricity through electromagnetic induction, where the movement of a conductor within a magnetic field generates an electric current.

## Energy Transmission

The electricity produced is then delivered over power lines to homes, companies, and industries, providing a steady and consistent supply of power.

## Conversion Mechanism

Hydro energy, also known as hydropower, captures the kinetic energy generated by flowing water to produce electricity. This conversion of energy can be achieved through various systems that utilize the natural movement of water found in rivers, streams, or artificial structures, such as dams and reservoirs. The basic principle is to convert the mechanical energy generated by flowing or falling water into electrical energy. There are several key mechanisms through which this conversion takes place:

## Hydroelectric Dams

Hydroelectric dams are the most common method for generating hydro energy. In this system, water is stored in a reservoir behind a dam and released through controlled gates. As the water flows from the higher elevation of the reservoir to a lower elevation, it passes through turbines. The potential energy held by the water at a higher elevation is converted into kinetic energy as it moves through the turbine, resulting in its rotation [65]. This spinning motion of the turbine powers a generator, which then generates electricity. The process involves:

- **Water Intake:** Water is drawn from the reservoir through large pipes called penstocks.
- **Turbine Activation:** The flowing water spins the turbine blades.
- **Energy Conversion:** The turbine is linked to a generator, which converts the mechanical energy of the rotating turbine into electrical energy by electromagnetic induction.
- **Discharge:** After passing through the turbine, the water is released downstream, continuing its natural course.

## Run-of-River Systems

Run-of-river hydroelectric systems create energy by utilising the natural flow of rivers, eliminating the need for massive reservoirs. They transfer a part of the river's water through a canal or penstock to spin a turbine before returning the water to the river [66]. Because they do not involve large dams or reservoirs, run-of-river systems are less disruptive to local ecosystems and are ideal for regions with consistent water flow. The conversion process is like that of dam-based systems, but the water is taken directly from the river without significant storage:

- **Water Diversion:** A part of the river’s flow is redirected through a penstock.
- **Turbine and Generator:** The water’s natural flow spins the turbine connected to the generator, converting mechanical energy into electricity.

- **Minimal Storage:** Unlike dam-based systems, run-of-river systems use little or no water storage, making them dependent on natural river flow rates.

### Pumped Storage Hydropower

Pumped storage is a unique form of hydro energy used primarily for energy storage rather than direct generation. The system consists of two reservoirs situated at varying heights. During times of reduced electricity demand, surplus electrical energy from the grid is utilized to pump water from the lower reservoir to the upper one. When electricity demand surges, the stored water is then released back to the lower reservoir through turbines, producing electricity. The pumped storage process involves:

- **Energy Absorption:** Water is pumped to the upper reservoir using surplus electricity during off-peak hours [67].
- **Energy Generation:** During peak demand, water is released from the higher reservoir to the lower reservoir, spinning turbines and generating energy.
- **Energy Storage:** This system essentially acts as a large battery, storing potential energy in the form of water at an elevated height and releasing it as needed.

### Tidal and Wave Energy

While not as prevalent, tidal and wave energy are types of hydro energy that utilize the movement of ocean waters to produce electricity. Tidal energy exploits the fluctuations in ocean tides to turn turbines, whereas wave energy harnesses the kinetic energy generated by surface waves.

- **Tidal Barrages:** These systems use barriers built across tidal estuaries. As the tide rises and falls, water flows through turbines in the barrage, generating electricity.
- **Tidal Stream Generators:** These devices are placed in fast-flowing tidal currents to capture the kinetic energy of moving water in a similar manner to wind turbines [68].
- **Wave Energy Converters:** These devices float on the surface of the water and move with the waves, using the motion to generate electricity.

Both tidal and wave energy are promising, though they are still in developmental stages due to technological challenges and environmental concerns.

## Micro-Hydropower Systems

Micro-hydropower systems are small-scale versions of hydroelectric power plants, typically designed to generate electricity for individual homes, small communities, or remote areas. These systems often utilize small streams or rivers to turn a turbine, providing a decentralized and renewable energy source.

- **Water Flow:** A tiny percentage of the stream's flow is redirected *via* a penstock.
- **Turbine Operation:** The water spins a small turbine, which is connected to a generator.
- **Localized Power Generation:** These systems generate relatively small amounts of electricity but are ideal for areas without access to a centralized power grid.

Micro-hydropower systems are cost-effective and environmentally friendly, requiring minimal infrastructure and causing little disruption to natural waterways.

### Advantages and Limitations

#### *Advantages of Hydro Energy*

- **Renewable and Sustainable:** Hydropower is considered a renewable energy source because it utilizes the natural water cycle, which includes precipitation and river flow, to produce electricity. If there is water, hydropower can continue to produce energy without depletion of resources.
- **Low Greenhouse Gas Emissions:** Compared to fossil fuels, hydropower plants generate minimal greenhouse gases once operational. This makes it an environmentally friendly option that helps mitigate climate change.
- **Long Lifespan:** Hydroelectric plants typically have extended operational lifetimes, often ranging from 50 to 100 years when adequately maintained. This longevity contributes to their cost-effectiveness over time and provides a reliable source of energy.
- **Flood Control and Water Management:** Many hydropower dams also serve additional purposes, such as controlling floods, providing irrigation for agriculture, and supporting local water management needs.
- **Job Creation:** The building, functioning, and upkeep of hydropower facilities generate employment opportunities, particularly in rural or isolated regions, thereby fostering economic growth in these communities.

#### *Limitations of Hydro Energy*

- **Displacement of Communities:** Large hydropower projects may necessitate the relocation of villages living along the river or reservoir. This displacement can cause social and cultural disruptions, as people may lose their homes, land, and livelihoods.

- **Dependence on Water Availability:** Hydropower generation depends on the availability of water. In regions affected by drought or changing precipitation patterns due to climate change, the efficiency and reliability of hydropower can be compromised.
- **Risk of Dam Failure:** Though rare, dam failures can have catastrophic consequences, including loss of life, property damage, and environmental destruction. Deteriorating or unfit maintenance can enhance dangers in the construction of hydropower dams.
- **Limited Geographic Suitability:** Hydropower is reliant on geography in the sense that development can be done only in areas where the water flow and elevation are adequate. This makes it very difficult to be adopted in most cases, especially in areas that do not have huge rivers or suitable land shapes.
- **Sedimentation:** Eventually, reservoirs are prone to sedimentation – this decreases the volume of water stored and the effectiveness of power generation. This condition needs regular maintenance using even capital-intensive dredging to remove accumulations, which also have adverse effects on the environment.

## Current Advancements and Implementations

Hydropower, as one of the oldest renewable energy sources, is being enhanced and is expanding in its ways of implementation as more and more power is being demanded while the environmental conscience becomes stronger. The focus of these improvements is on increasing the effectiveness, lowering the ecological impact, and making hydro energy usable in new ways.

## Upgrades to Existing Infrastructure

The improvement of existing hydropower infrastructure has tremendously enhanced both efficiency and capacity. Upgrades include installation of advanced turbines, water flow optimization, and modern control systems. These improvements increase energy production without the construction of new dams, which reduces the negative impact on nature.

## Pumped Storage Hydropower (PSH)

Pumped strategies have become ubiquitous globally and are expected to proliferate lateral bi-national power systems throughout the world. This method frequently refers to a system that overcomes “negative energy prices.” Many technological advancements incorporate closed-loop “PSH” systems, which reduce environmental impact and expand applicability to more regions [69].

## Hydropower-Integrated Smart Grids

Because of the intermittent nature of renewable energy sources like wind and solar, grid management has become more complex. Pumped hydro storage, and particularly Pumped Storage Plants (PSPs), are integrated into smart grids for energy supply and demand management. Surplus solar and wind energy is stored with these systems, which utilize sophisticated sensors, automation, and real-time analysis to optimize operations and curtail reliable power delivery.

## Floating Hydropower

Floating hydropower systems consist of turbines mounted on floating structures, which maintain power generation in places with seasonal flooding or fluctuating water depths. This new technique is environmentally friendly and lowers the cost of construction as compared to conventional dams: it is thus a viable option for areas where rivers' level changes widely.

## Hydropower in Hybrid Systems

Today, hydropower is being incorporated into hybrid renewable energy systems along with other renewable sources like wind, solar or geothermal energy sources. It stabilizes energy by interfacing the non-firm output of wind and solar energy with the firm output of hydropower. To enhance efficiency, floating solar panels on hydropower reservoirs also minimize space and evaporation of water.

## Future Trends and Innovations

Hydropower is one of the oldest and most reliable forms of renewable energy. And despite being a skilled and matured industry, changes in hydropower development are taking place to suit current energy needs and issues of the environment. Given the global situation around decarbonization, it can be assumed that future trends and innovations in hydro energy are expected to focus on enhancing efficiency, reducing ecological impacts, and integrating with emerging technologies. The following are some important trends and breakthroughs influencing the future of hydro energy:

## Digitalization and Smart Hydropower

The future application of digitalization principles in hydropower plants is inspirational. Advanced monitoring systems equipped with Artificial Intelligence and the Internet of Things will soon enable the collection of real-time data and conduct predictive maintenance of the facilities and systems, thus reassuring the optimized operation of the plant. These smart hydropower systems can improve efficiency, reduce downtime, and extend the lifespan of equipment [70].

Moreover, digitalization enables better water management, ensuring hydropower plants adapt to changing water availability and climate conditions.

### Hydrokinetic Energy and Marine Hydropower

The use of hydrokinetic energy, obtained from the forces of moving water from natural rivers, tides, and ocean currents, is emerging as a more environmentally sustainable energy source. Future innovations seek to enhance the turbine's efficacy and hardiness in aquatic environments. Tidal and wave energy converters possess the capacity to harness kinetic energy in oceans, which has the potential to benefit coastal communities. Economical and reliable marine hydropower systems could also enhance the hydropower fisheries globally, where it was the only resource.

### Hybrid Hydropower Systems

Harnessing renewable energy sources in hybrid power plants is becoming popular; therefore, plants that use energy sources with complementary generation patterns are built. For instance, wind turbines and solar panels can be utilized on sunny or windy days and hydropower can be used when the solar output is lower. This, in turn, decreases reliance on fossil fuels and stabilizes the grid.

### Energy Market Integration

As the world is progressing towards globalization, it has become very clear that with the use of energy that is variable in nature, particularly wind and solar energy, hydropower is a need of the hour to provide stability of supply [71]. As time goes on, hydropower stations will become more and more active in ancillary services such as Demand Response and Frequency Regulation, which will be beneficial for the stability of the grid.

### Retrofitting and Upgrading Existing Infrastructure

The next step in the development of hydropower will not be the construction of new large structures, but their reconstruction. For the replacement of aging dams, more and more turbines with several hundred years of patent life will be required, and automation systems and water management technology upgrades are already available. Hydropower can be further developed by adding hydropower features to current dams rather than building new dams, which avoids environmental impacts and lowers costs.

While hydro energy capitalizes on water flow, biomass energy converts organic waste into power, representing a critical intersection between energy and sustainable waste management.

# BIOMASS ENERGY

## Fundamental Principle

Biomass energy comes from biological materials, or the organic, industrious part of life that comes from plant and animal matter. The essence of biomass energy is in its harvesting of materials that have captured solar energy or rather, half the biomass energy is available as solar energy. Solar energy is captured through photosynthesis good leaves that turn water and carbon dioxide into glucose and oxygen with the help of the sun. This energy can be harnessed and utilized in different forms by breaking chemical bonds in plant tissues. When biomass materials, such as wood, agricultural residues, or animal waste, are burned or otherwise processed, the chemical energy stored in their organic matter is released as heat [56]. This heat can be utilised directly for heating or converted into electricity using steam turbines. Alternatively, biomass can be transformed into biofuels, such as ethanol or biodiesel, through biochemical or thermochemical processes. These biofuels can be utilised as alternatives to fossil fuels in transportation and other purposes. In essence, biomass energy capitalizes on the sun's energy that plants have captured over time.

By converting this stored solar energy, biomass serves as a renewable energy source, helping to reduce reliance on fossil fuels and lower greenhouse gas emissions.

## Conversion Mechanism

Biomass energy comes from organic materials like plant matter, agricultural residues, and animal waste. As a renewable energy source, it can be transformed into various usable energy forms, including electricity, heat, and biofuels. The conversion of biomass into energy typically involves one of three primary mechanisms: thermal conversion, chemical conversion, and biological conversion.

## Thermal Conversion

Thermal conversion processes use heat to transform biomass into energy through methods like combustion, pyrolysis, and gasification, as shown in Fig. (**6**).

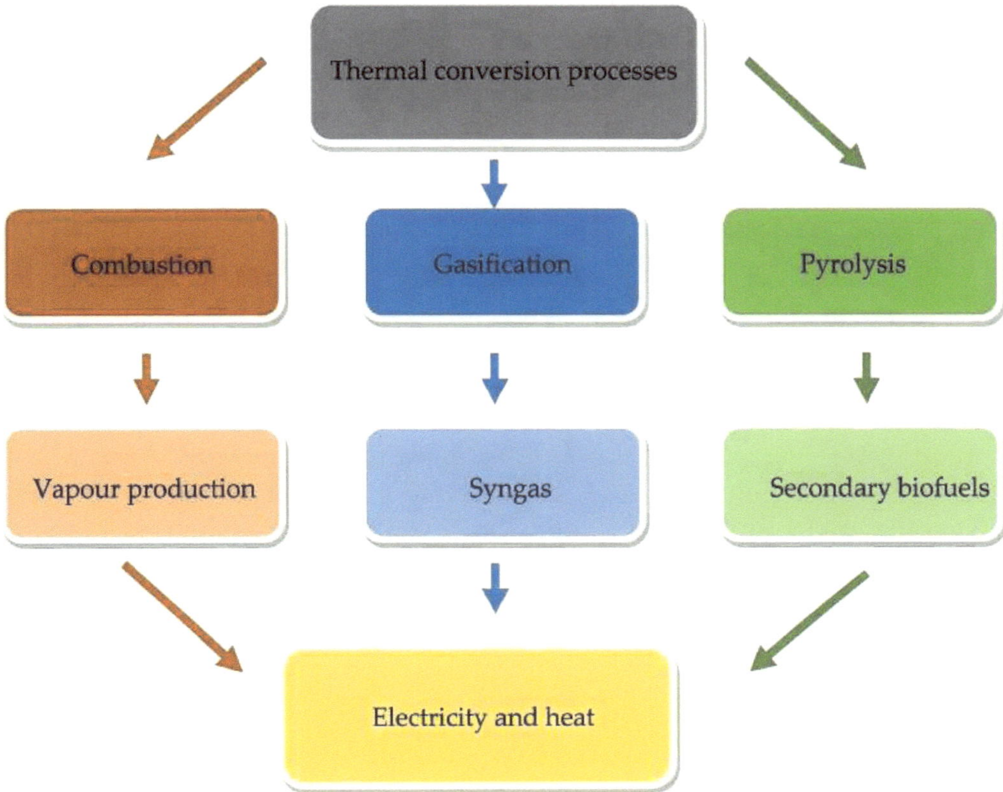


**Fig.(6).** Mechanism of thermal conversion [72].

- **Combustion:** This is the most straightforward and widely used method for biomass energy production [73]. Biomass is combusted with oxygen to produce heat, which can be used directly for heating or to generate steam to power turbines for electricity production. Modern biomass combustion systems are designed to improve efficiency and reduce emissions compared to traditional methods like open burning.
- **Pyrolysis:** Pyrolysis is a process that heats biomass in an oxygen-free environment, leading to its breakdown into three primary products: bio-oil (liquid), syngas (gas), and biochar (solid). Each of these products has various potential uses—bio-oil can be further processed into fuels, syngas can be utilized for power generation or as a raw material for chemicals, and biochar can act as a soil enhancer or a means of carbon sequestration.
- **Gasification:** In gasification, biomass is converted into a combustible gas mixture known as syngas, primarily consisting of carbon monoxide, hydrogen, and methane. This process occurs at high temperatures with limited oxygen.

Syngas can be utilized for electricity generation, heating purposes, or as a raw material for creating synthetic fuels.

### Chemical Conversion

Chemical conversion processes transform biomass into fuels and chemicals using catalysts and chemical reactions. Two primary methods are transesterification and hydrothermal liquefaction.

- **Transesterification:** This chemical process is primarily used to convert oils and fats from biomass into biodiesel. The procedure involves mixing the oil with alcohol, typically methanol, using a catalyst. This reaction results in the formation of biodiesel, along with glycerol as a byproduct. Biodiesel serves as a renewable energy source that can either substitute for or be blended with the traditional diesel fuel.
- **Hydrothermal Liquefaction:** In this process, biomass is subjected to water at higher temperatures and pressures, breaking down the organic material into bio-crude oil. This bio-crude can be refined into liquid fuels like petroleum-based products, making it a promising alternative for transportation fuels.

### Biological Conversion

Biological conversion relies on microorganisms to convert organic matter into energy-rich fuels. This process includes anaerobic digestion and fermentation.

- **Anaerobic Digestion:** Anaerobic digestion occurs when microorganisms break down organic material, such as agricultural waste or food waste, in the absence of oxygen [74]. This process produces biogas, a mixture of methane and carbon dioxide, which can be captured and used for electricity generation, heating, or as a renewable natural gas. The leftover solid material can be used as a fertilizer.
- **Fermentation:** In fermentation, microorganisms, usually yeast or bacteria, transform carbohydrates into biomass, ethanol, or other biofuels. This process is commonly used to produce ethanol from crops like corn, sugarcane, or cellulosic biomass.

### Hybrid Conversion Processes

In addition to these primary conversion mechanisms, hybrid approaches that combine elements of thermal, chemical, and biological processes are being developed. These systems aim to maximize energy efficiency, minimize waste, and create versatile energy products.

## Advantages and Limitations

### *Advantages of Biomass Energy*

- **Renewable Resource:** Biomass energy comes from organic substances, including plants, agricultural byproducts, and waste materials. If these resources are sustainably managed, biomass can be replenished, making it a renewable energy source.
- **Carbon Neutral Potential:** Biomass energy is often regarded as a potentially carbon-neutral source. The combustion of biomass emits $CO_2$ gas, which, during photosynthesis, green plants can capture, thus achieving a balance. Proper management of this process helps to reduce the net growth of $CO_2$ in the atmosphere, enabling the reversal of climate change effects.
- **Energy Security:** It is possible to enhance biomass energy generation locally, which means there is less dependence on imported fossil fuels. Such a policy would improve energy independence and generate economic viability, especially in regions with surplus biomass material.
- **Versatility in Applications:** Biomass can be used for the generation of electricity and heat, as well as liquid fuels, thus it is a flexible source of energy. For instance, biomass can be used to produce biofuels like ethanol and biodiesel, which are alternative fuels for automobiles.
- **Supports Rural Economies:** Biomass production and its primary and secondary processing can create jobs for rural communities that are dominated by farming and forestry work. This would, in turn, motivate the local markets and increase income for landowners and farmers.

### *Limitations of Biomass Energy*

- **Resource Intensity:** The growth, harvesting, and processing of biomass materials and their end-products require large amounts of land, water, and energy resources. The extensive expansion of biomass use could lead to competition with food crops for limited and scarce arable land and water resources, raising concerns about food supply.
- **Carbon Emissions and Air Pollution:** In principle, biomass energy is believed to be carbon neutral; however, there are still possibilities for its generation to emit greenhouse gases and air pollutants if it is not sustainably managed [75]. Particulate matter, $NO_x$, and other gaseous species are released due to the incomplete combustion of biomass, which can contribute to air quality issues.
- **Land Use and Deforestation:** The increased demand for biomass resources can result in deforestation, habitat loss, and soil deterioration if forests and natural ecosystems are converted into biomass plantations. Sustainable management practices are essential to avoid environmental damage.

- **High Initial Costs:** Establishing biomass energy plants and infrastructure can involve high upfront costs, including the purchase of specialized equipment and technology. Furthermore, since biomass materials are energy dense, moving them over long distances poses significant economic challenges.
- **Seasonal and Geographic Limitations:** Biomass availability may vary and may be affected by seasonal as well as geographical characteristics. Certain regions may also have limited biomass resources, which may affect the level of biomass energy production in those regions.

### Current Advancements and Implementations

Energy derived from biomass, which is the source of organic matter like wood and agricultural residues, has shown significant changes over the years. These advances enhance the potential of biomass as a renewable energy source, its efficiency, and market adoption. Focus on the reduction of carbon emissions and increasing energy security has also prompted the emergence of different technologies and practices in the biomass energy space.

### Bioenergy with Carbon Capture and Storage (BECCS)

BECCS is already proving and is bound to achieve significant reductions in carbon dioxide in the atmosphere, as it supports environmental safety. BECCS is the technique of capturing $CO_2$ produced by the combustion or processing of biomass. The carbon that has been captured can be used in various industries or stored underground. This approach not only generates energy but also removes $CO_2$ from the atmosphere, resulting in negative emissions.

Several pilot projects are currently exploring the feasibility of BECCS, with potential applications in large-scale biomass power plants and biofuel production facilities.

### Second-Generation Biofuels

Second-generation biofuels are derived from non-food biomass sources, addressing concerns about competition between biofuel production and food supply. These fuels are derived from agricultural byproducts, forestry residues, and dedicated energy crops such as switchgrass and algae. Advances in lignocellulosic biofuel technology, which transforms sturdy plant fibres into usable fuel, are making biofuels more efficient and affordable. Algal biofuels are receiving attention for their high yield potential and ability to grow in non-arable land using saline water [76].

## Waste-to-Energy Technologies

Innovative waste-to-energy technologies are transforming organic waste into useful energy. Municipal solid waste, agricultural residues, and industrial byproducts can be converted into biogas, electricity, or heat, utilising techniques such as anaerobic digestion and incineration. Advanced sorting and processing technologies are enhancing the ability to extract biomass from waste streams, maximizing energy recovery while reducing landfill use. Gasification of waste is also becoming more efficient, with advancements in reactor design and feedstock flexibility.

## Combined Heat and Power (CHP) Systems

Biomass-powered Combined Heat and Power (CHP) systems generate both electricity and useful heat from a single fuel source, maximizing energy efficiency. These systems are being widely implemented in industries, particularly in the paper, pulp, and food processing sectors, where they can utilize biomass waste products. Moreover, advances in small-scale CHP systems have made them easier to use in residential and commercial sectors, allowing for more decentralized renewable energy generation.

## International Implementations

Countries worldwide are progressively allocating resources to biomass energy as a component of their renewable energy strategies. Some noteworthy implementations include:

- **Europe:** The European Union has been a leader in biomass energy, with countries like Sweden, Finland, and the UK relying heavily on biomass for power generation and heating. In Denmark, biomass-based CHP plants provide district heating to entire cities. The EU's Renewable Energy Directive has also set ambitious targets for bioenergy use [77].
- **United States:** In the U.S., biomass energy is widely used in the form of wood pellets, biogas from landfills, and ethanol from corn. Large-scale biomass power plants and biofuel facilities are operating across the country, particularly in regions with abundant forestry or agricultural waste.
- **Developing Countries:** In regions like Sub-Saharan Africa and Southeast Asia, biomass energy plays an important role in rural electrification and clean cooking solutions. Recent projects involve the deployment of small-scale biogas digesters and biomass gasification units to provide sustainable energy access in off-grid communities.

## Future Trends and Innovations

For decades, biomass energy has been a valuable renewable energy source derived from organic materials such as plants, agricultural wastes, and garbage. As the world continues to seek sustainable alternatives to fossil fuels, biomass energy is poised for significant advancements and innovations. Biomass energy is bound to be affected just like any other industry today due to changes in technology, the rise of efficiency, and the adoption of other renewable energy enterprises. Let us investigate some of the most promising biomass energy developments and breakthroughs that will change the future of this energy completely:

### Biomass Gasification and Syngas Production

Biomass gasification, which seeks to transform organic matter/natural materials into Syngas, a gas consisting of hydrogen, carbon monoxide, and carbon dioxide, has increasingly become a popular way of generating clean energy. The syngas created can be gasified further to produce liquid biofuels or burned in gas turbines to generate energy. One such process, which is being perfected, involves plasma and supercritical water, offering greater efficiency and managing lower emissions than traditional gasification processes. In the future, biomass gasification is likely to become commonplace within the waste-to-energy systems as it has the potential to dispose of waste while also producing energy sustainably. This trend also offers promise for green hydrogen production, an energy carrier for the future, which can also be obtained from biomass gasification.

### Integrated Biomass and Renewable Energy Systems

Future plans are to operate biomass energy systems alongside solar and wind energy systems to form hybrid energy systems. This will produce more stable and secure electricity and enable generation practitioners to be geographically dispersed in locations where sunlight and wind energy are not always available. For instance, when solar or wind energy is not being harnessed, biomass energy can be used to replace them as a supplemental energy source. In some countries, hybrid systems combining biomass and other sustainable sources are already being implemented successfully and proving to be cost-effective and environmentally sustainable. The synergy between biomass and other renewables will improve the overall energy efficiency and decrease dependence on fossil fuels.

### Improved Biomass Conversion Technologies

Technological developments in thermochemical and biochemical conversion processes will be key in increasing the efficacy of biomass energy generation from biomass. Such innovations include pyrolysis, hydrothermal liquefaction, fermentation techniques, *etc.* If pyrolysis refers to thermal decomposition of organic material without oxygen, hydrothermal liquefaction is a process turning wet biomass into bio-crude and fermentation helps to process biomass into biofuels, biochar, and many more useful products. These improved conversion technologies will assist in minimizing the energy losses during biomass-to-energy process and enhance increase in energy output making biomass energy more competitive amongst other renewable energy sources. Further development of anaerobic digestion, which transforms organic waste into biogas, should also facilitate the use of this technology for energy generation in both rural and urban settings.

## Sustainable Biomass Feedstock Sourcing

Biomass feedstock procurement will dominate the future of biomass energy as cutting-edge technology will empower biomass producers, processors, and manufacturers to deal with issues of deforestation and land use competition [78]. Innovations in feedstock production are expected to revolve around the use of marginal lands, crop residues, and waste-to-energy systems. In addition, research is underway to propagate energy crops that require low inputs and can survive in harsh conditions, and thus lessen the ecological impacts of biomass production. Furthermore, the cultivation of algae, seaweed, and other forms of aquatic biomass is being promoted as they grow faster and do not need suitable agricultural land [79]. These feedstocks could provide a cheaper and environmentally friendly biomass for power generation in the future without impeding food production.

## AI and Digitalization in Biomass Energy

Biomass energy production and control will not be the same again with the introduction of AI and digital advancement. Biomass fuel supply chains can be optimized, energy demands can be predicted, and biomass conversion processes improved with AI. Blockchains, smart sensors, and machine learning are some of the modern technologies that will allow monitoring and enhancement of biomass systems in real time. These technologies will also facilitate more precise and efficient management of biomass resources, ensuring that biomass feedstocks are used in the most sustainable and economically viable manner.

The following Table **1** summarizes the key economic and operational metrics across the major renewable energy technologies discussed above, highlighting their comparative feasibility.

**Table 1. Cost and efficiency comparison of renewable energy technologies.**

| Technology | Initial Cost ($/kW) | O&M Cost ($/kWh) | LCOE ($/kWh) | ROI / Payback Period (Years) | Subsidies / Incentives |
|---|---|---|---|---|---|
| Solar PV | 900 - 1,400 | 0.005 - 0.02 | 0.03 - 0.06 | 5 - 8 | Tax credits, feed-in tariffs, green loans |
| Wind (Onshore) | 1,200 -1,700 | 0.007 - 0.02 | 0.03 - 0.05 | 6 - 10 | Production tax credits, green bonds |
| Hydropower | 1,000 -3,500 | 0.005 - 0.015 | 0.03 - 0.07 | 10 - 20 | Long-term PPA support, grants |
| Geothermal | 2,500 -5,000 | 0.01 -0.03 | 0.04 - 0.10 | 10 -15 | Risk insurance, drilling subsidies |
| Biomass | 1,500 -3,000 | 0.02 – 0.04 | 0.06 – 0.11 | 8 - 12 | Feed-in tariffs, tax relief on fuel |

As shown, solar and wind technologies have the Lowest Levelized Cost Of Electricity (LCOE), with relatively quick Returns on Investment (ROI), especially where incentives are strong. Geothermal and hydropower, while more capital-intensive, offer high-capacity factors. Biomass generally presents higher ongoing costs and variable ROI, often dependent on feedstock availability and local policy support.

Having discussed each renewable technology in detail, the following comparative analysis synthesizes their relative strengths, environmental impacts, and economic viability.

## COMPARATIVE ANALYSIS

### Efficiency of Various Technologies

**Solar Energy:** The efficiency of Photovoltaic (PV) solar panels typically ranges from 15% to 22%, with newer technologies such as perovskite cells promising higher efficiencies [80]. The efficiency of solar energy is heavily impacted by factors such as geographical location, weather patterns, and the angle at which sunlight strikes the surface. Solar energy has vast use, but its effectiveness can be limited by the intermittent nature of sunshine and the demand for extensive surface areas.

**Wind Energy:** Modern wind turbines operate with efficiencies ranging between 35% and 45%, making wind energy one of the more efficient renewable sources. Wind efficiency is influenced by the turbine's size, design, and wind speed.

Nonetheless, the generation of wind energy is also subject to intermittency, as it depends on the presence of robust and steady wind currents.

**Geothermal Energy:** Geothermal energy boasts the highest efficiency among renewable sources, with systems achieving up to 90% efficiency. Unlike solar and wind, geothermal energy can provide continuous power generation since it relies on the earth's internal heat, which is not subject to daily or seasonal variations [81].

**Hydro Energy:** Hydropower has a high conversion efficiency, generally ranging from 70% to 90%. It benefits from steady, predictable water flows, particularly in dam-based systems, which allow for flexible control over electricity generation. However, its efficiency can decrease during droughts or periods of low water availability.

**Biomass Energy:** Biomass conversion efficiency varies depending on the technology used. Direct combustion of biomass for electricity generation typically has an efficiency of 20% to 30%, while advanced techniques like gasification or co-firing with other fuels can achieve higher efficiencies, approaching 35% to 40%. Biomass efficiency depends on the type of feedstock and the technology used in conversion.

**Environmental Impacts**

**Solar Energy:** Solar power is a clean and renewable energy source with minimal direct environmental impacts during operation [82]. However, the production of solar panels involves mining and manufacturing processes that generate emissions, and end-of-life disposal can lead to waste challenges. Large solar farms may also disrupt local ecosystems by occupying a significant land area.

**Wind Energy:** Wind power produces no emissions during operation, making it environmentally friendly. The most notable environmental concerns are bird and bat mortality due to collisions with turbines and the potential visual and noise impacts on nearby communities. The land use of wind farms can also lead to habitat fragmentation, although agricultural or other activities can often coexist with wind turbines.

**Geothermal Energy:** Geothermal energy has a relatively low environmental footprint, as it produces minimal emissions during operation. The extraction of geothermal fluids may lead to the release of greenhouse gases that are trapped below the Earth's surface. Additionally, there are concerns regarding the potential for land subsidence and the contamination of water sources [83]. Still, these impacts are generally localized and can be managed with proper regulation.

**Hydro Energy:** Hydropower has significant environmental implications, especially when dams are involved. The building of large dams can cause the destruction of natural habitats, displace local populations, and disturb aquatic ecosystems. Dams can modify the flow of rivers, interfere with fish migration patterns, and impact the quality of water in downstream areas. While small-scale hydropower projects have fewer environmental impacts, large-scale systems require careful management to mitigate damage.

**Biomass Energy:** The environmental impacts of biomass energy vary depending on the feedstock used [84]. Although biomass is theoretically considered carbon-neutral—since the carbon dioxide released during combustion is balanced by the amount absorbed during plant growth—real-world practices like cultivation, harvesting, and transportation can lead to additional emissions. Deforestation, land-use changes, and air pollution from burning certain types of biomass can also contribute to environmental degradation.

## Economic Viability

Economic viability refers to the cost-effectiveness and long-term sustainability of a technology [85], typically evaluated using metrics such as Return on Investment (ROI), Levelized Cost of Electricity (LCOE), payback period, and the impact of financial mechanisms such as subsidies and tax credits. This section assesses the economic viability of each major renewable energy source, and the comparison of different energy sources in terms of efficiency, environmental impact, and economic viability is given in Table **2**.

**Solar Energy:** Solar energy has become more economically viable in recent years due to falling prices of solar panels and advances in technology [86]. Initial installation costs can be high, but government incentives and subsidies help offset this, and operational costs are low due to the free availability of sunlight. Solar power is particularly cost-effective in regions with high solar irradiance [87]. However, energy storage solutions, such as batteries, are still relatively expensive, which can affect the economic feasibility of solar energy for a continuous power supply.

**Wind Energy:** Wind energy has achieved strong economic viability, especially in areas with consistent wind resources. Installation costs for wind turbines have dropped significantly, making wind power one of the most cost-competitive renewable energy sources [88]. Like solar, wind energy has low operational costs after installation. However, initial capital investment is still substantial, and transmission infrastructure may be needed to deliver electricity from remote wind farms to population centres.

**Geothermal Energy:** Geothermal power plants require substantial upfront investment due to the need for drilling and exploration to tap into geothermal reservoirs. However, once established, geothermal energy is highly cost-effective, providing stable and continuous energy with minimal fuel costs. Geothermal energy can have lower long-term costs compared to other renewables, especially in regions with rich geothermal resources. Still, its use is restricted to areas where geothermal heat is readily available.

**Hydro Energy:** Hydropower stands out as one of the most affordable renewable energy options, particularly for large-scale installations. The long lifespan of hydropower plants and low operational costs contribute to their economic advantage. However, the high capital costs associated with dam construction, environmental mitigation, and potential social costs *(e.g)*, displacement of communities) can be prohibitive for new projects. Small-scale hydropower is more cost-effective and can be installed in a broader variety of sites, while causing less environmental harm.

**Biomass Energy:** It has mixed economic viability. The cost of biomass feedstock can vary, and transporting large volumes of material can add to expenses. Additionally, the infrastructure for biomass energy production, such as power plants, requires significant investment. In certain regions where biomass is abundant and transport distances are short, biomass energy can be cost-effective. However, its overall competitiveness depends on regional factors, feedstock availability, and the specific technology used.

**Table 2. Comparison of Energy Sources: Efficiency, Environmental Impact, and Economic Viability.**

| Energy Source | Efficiency | Environmental Impact | Economic Viability |
|---|---|---|---|
| Solar | 15%–22% (higher with new tech) | Minimal emissions, land use, and panel disposal can occur. | High initial costs; increasingly competitive in sunny areas. |
| Wind | 35%–45% | Minimal emissions; visual, noise, and wildlife impacts. | Competitive; reliant on wind consistency. |
| Geothermal | 70%–90% | Low emissions; potential for water and land issues. | High upfront cost; very cost-effective in geothermal-rich areas. |
| Hydro | 70%–90% | It can disrupt ecosystems and water quality. | Cost-effective but expensive to build large dams. |
| Biomass | 20%–40% | Carbon-neutral if managed well; risk of air pollution and deforestation. | Viable where biomass is abundant; variable feedstock costs. |

## CONCLUSION

Renewable energy technologies are essential in the global transition to sustainable energy systems. In response to the environmental, economic, and social issues caused by traditional fossil fuel reliance, alternatives such as solar, wind, geothermal, hydro, and biomass energy offer cleaner and more sustainable options. While each technology presents its own advantages and challenges, together they offer versatile solutions that can be adapted to various geographic regions, economic conditions, and energy requirements. With its vast availability and continuous efficiency improvements, solar energy is revolutionizing how we capture sunlight to generate electricity and heat. Wind energy, especially in regions with consistent wind patterns, is becoming a cornerstone of many countries' renewable energy portfolios. Geothermal energy harnesses the Earth's internal heat, providing a dependable and steady power source, especially in regions with advantageous geological features. Hydro energy, one of the most established renewable technologies, continues to provide large-scale energy solutions, while newer innovations in small-scale hydro and tidal energy are expanding its potential. Biomass energy, utilizing organic materials, is unique in its ability to convert waste into a valuable resource, contributing to both energy generation and waste management. Despite their benefits, these technologies face certain challenges, such as intermittency, geographic limitations, and the need for significant infrastructure investments. Ongoing research and advancements in technology are working to overcome these challenges, leading to the creation of more efficient, affordable, and integrated energy systems. The future of renewable energy will be shaped by further progress in hybrid systems, improved energy storage technologies, and forward-thinking policies that encourage broader adoption. In the global effort to lower carbon emissions and address the challenges of climate change, renewable energy technologies play an increasingly critical role. By expanding the range of energy sources and shifting toward sustainable solutions, we can build a cleaner, more robust energy future that supports both environmental protection and economic development while enhancing energy security.

To accelerate this transition, governments, industries, and research institutions must adopt supportive policy frameworks, increase investment in advanced renewable technologies, and implement financial mechanisms such as subsidies and tax incentives. Strategic infrastructure planning, public-private partnerships, and effective community engagement will also be essential in scaling up adoption. Without decisive action today, the opportunity to mitigate the climate crisis and ensure equitable energy access for future generations may be irreversibly lost.

# REFERENCES

[1] I. Dincer, "Renewable energy and sustainable development: a crucial review", *Renew. Sustain. Energy Rev.,* vol. 4, no. 2, pp. 157-175, 2000.
[http://dx.doi.org/10.1016/S1364-0321(99)00011-8]

[2] R. Kumar, M. Sethi, V. Goel, M.K. Ramis, M. AlSubih, S. Islam, M.S. Al Ansari, D. Lee, and A. WogassoWodajo, "Thermal and effective assessment of solar thermal air collector with roughened absorber surface: an analytical examination", *Int. J. Low Carbon Technol.,* vol. 19, pp. 1112-1123, 2024.
[http://dx.doi.org/10.1093/ijlct/ctae056]

[3] O. Ellabban, H. Abu-Rub, and F. Blaabjerg, "Renewable energy resources: Current status, future prospects and their enabling technology", *Renew. Sustain. Energy Rev.,* vol. 39, pp. 748-764, 2014.
[http://dx.doi.org/10.1016/j.rser.2014.07.113]

[4] K. Kumar, S. Kumar, D. Prajapati, S. Samir, S. Thapa, and R. Kumar, "Solar thermal collector roughened with s-shaped ribs: parametric optimization using AHP-MABAC technique", *Fluids,* vol. 10, no. 3, p. 67, 2025.
[http://dx.doi.org/10.3390/fluids10030067]

[5] A. Demirbas, "Global renewable energy projections", *Energy Sources B Econ. Plan. Policy,* vol. 4, no. 2, pp. 212-224, 2009.
[http://dx.doi.org/10.1080/15567240701620499]

[6] R. Kumar, M. Sethi, A. Thakur, A. Rana, V. Goel, D. Lee, and T. Singh, "Analysis of exergy in a dimple-roughened solar thermal collector using MATLAB simulation", *J. Therm. Anal. Calorim.,* vol. 150, no. 1, pp. 433-449, 2025.
[http://dx.doi.org/10.1007/s10973-024-13707-2]

[7] R. Kumar, N. Sharma, Chahat, G.H. Ngo, A.K. Yadav, D. Lee, and T. Singh, "Prediction of jet impingement solar thermal air collector thermohydraulic performance using soft computing techniques", *Case Stud. Therm. Eng.,* vol. 55, p. 104144, 2024.
[http://dx.doi.org/10.1016/j.csite.2024.104144]

[8] Y. Kuang, Y. Zhang, B. Zhou, C. Li, Y. Cao, L. Li, and L. Zeng, "A review of renewable energy utilization in islands", *Renew. Sustain. Energy Rev.,* vol. 59, pp. 504-513, 2016.
[http://dx.doi.org/10.1016/j.rser.2016.01.014]

[9] M.K. Singla, "Optimizing solar cell models: A multi-objective parameter estimation algorithm for improved photovoltaic system performance", *Phys. Scr.,* 2024.

[10] P.A. Østergaard, N. Duic, Y. Noorollahi, and S. Kalogirou, "Renewable energy for sustainable development", *Renew. Energy,* vol. 199, pp. 1145-1152, 2022.
[http://dx.doi.org/10.1016/j.renene.2022.09.065]

[11] H. Lund, "Renewable energy strategies for sustainable development", *Energy,* vol. 32, no. 6, pp. 912-919, 2007.
[http://dx.doi.org/10.1016/j.energy.2006.10.017]

[12] R. Thakur, T. Kashyap, R. Kumar, R.K. Saini, D. Lee, S. Kumar, and T. Singh, "Potential of the Archimedes screw to generate sustainable green energy for mini, micro, and pico hydro Turbine power stations: An extensive analysis", *Energy Strategy Reviews,* vol. 55, p. 101514, 2024.
[http://dx.doi.org/10.1016/j.esr.2024.101514]

[13] R. Kumar, S. Kumar, Ü. Ağbulut, A.E. Gürel, M. Alwetaishi, S. Shaik, C.A. Saleel, and D. Lee, "Parametric optimization of an impingement jet solar air heater for active green heating in buildings using hybrid CRITIC-COPRAS approach", *Int. J. Therm. Sci.,* vol. 197, p. 108760, 2024.
[http://dx.doi.org/10.1016/j.ijthermalsci.2023.108760]

[14] R. Avtar, N. Sahu, A.K. Aggarwal, S. Chakraborty, A. Kharrazi, A.P. Yunus, J. Dou, and T.A. Kurniawan, "Exploring renewable energy resources using remote sensing and GIS—A review",

*Resources,* vol. 8, no. 3, p. 149, 2019.
[http://dx.doi.org/10.3390/resources8030149]

[15] P. K. Pandey, Z. EL KHATIB, R. Kumar, M. Singla, M. Safaraliev, and F. Kamalov, "A novel hybrid control strategy for grid-connected PV-battery system under different operating conditions," Przegląd Elektrotechniczny, 2024.

[16] A. Garg and R. K. Ratnesh, "Solar cell trends, and the future: A review," Journal of Pharmaceutical Negative Results, vol. 13, pp. 2051-2060, 2022.
[http://dx.doi.org/10.47750/pnr.2022.13.S06.268]

[17] A.K. Yadav, V. Yadav, A. Kumar, R. Kumar, D. Lee, and T. Singh, "Novel feature selection based ANN for optimal solar panels tilt angles prediction in micro grid", *Case Stud. Therm. Eng.,* vol. 61, p. 104853, 2024.
[http://dx.doi.org/10.1016/j.csite.2024.104853]

[18] R. Kumar, S. Kumar, R. Nadda, K. Kumar, and V. Goel, "Thermo-hydraulic efficiency and correlation development of an indoor designed jet impingement solar thermal collector roughened with discrete multi-arc ribs", *Renew. Energy,* vol. 189, pp. 1259-1277, 2022.
[http://dx.doi.org/10.1016/j.renene.2022.03.037]

[19] A. Jamar, Z.A.A. Majid, W.H. Azmi, M. Norhafana, and A.A. Razak, "A review of water heating system for solar energy applications", *Int. Commun. Heat Mass Transf.,* vol. 76, pp. 178-187, 2016.
[http://dx.doi.org/10.1016/j.icheatmasstransfer.2016.05.028]

[20] S. Mekhilef, R. Saidur, and A. Safari, "A review on solar energy use in industries", *Renew. Sustain. Energy Rev.,* vol. 15, no. 4, pp. 1777-1790, 2011.
[http://dx.doi.org/10.1016/j.rser.2010.12.018]

[21] S. Kumar, R. Thakur, S. Kumar, D. Lee, and R. Kumar, "Impact of liquid spectrum filter and solar tracker on the overall effectiveness of a photovoltaic thermal system: An experimental investigation", *Renew. Energy,* vol. 226, p. 120390, 2024.
[http://dx.doi.org/10.1016/j.renene.2024.120390]

[22] S. Kumar, R. Kumar, R. Thakur, S. Kumar, and D. Lee, "Effects of climate variables and nanofluid-based cooling on the efficiency of a liquid spectrum filter-based concentrated photovoltaic thermal system", *J. Therm. Anal. Calorim.,* vol. 149, no. 5, pp. 2273-2291, 2024.
[http://dx.doi.org/10.1007/s10973-023-12836-4]

[23] P.K. Pandey, R. Kumar, V. Gupta, A. Kanungo, and S. Diwania, "An innovative hybrid controller-based combined grid-connected hybrid renewable energy system", *Electr. Eng.,* vol. 106, no. 6, pp. 7055-7072, 2024.
[http://dx.doi.org/10.1007/s00202-024-02363-2]

[24] V. Gupta, M. Sharma, R.K. Pachauri, and K.N. Dinesh Babu, "Comprehensive review on effect of dust on solar photovoltaic system and mitigation techniques", *Sol. Energy,* vol. 191, pp. 596-622, 2019.
[http://dx.doi.org/10.1016/j.solener.2019.08.079]

[25] J. Keane, *Pico-solar electric systems: The Earthscan expert guide to the technology and emerging market.* Routledge, 2014.
[http://dx.doi.org/10.4324/9781315818467]

[26] S. Aggarwal, R. Kumar, D. Lee, S. Kumar, and T. Singh, "A comprehensive review of techniques for increasing the efficiency of evacuated tube solar collectors", *Heliyon,* vol. 9, no. 4, p. e15185, 2023.
[http://dx.doi.org/10.1016/j.heliyon.2023.e15185] [PMID: 37089311]

[27] S. Aggarwal, R. Kumar, S. Kumar, and T. Singh, "Impact of fin material properties and the inclination angle on the thermal efficiency of evacuated tube solar water heater: An experimental study", *J. King Saud Univ. Sci.,* vol. 36, no. 5, p. 103186, 2024.
[http://dx.doi.org/10.1016/j.jksus.2024.103186]

[28] R. Kumar, R. Nadda, S. Kumar, S. Saboor, C. Ahamed Saleel, M. Abbas, A. Afzal, and E. Linul,

"Convective heat transfer enhancement using impingement jets in channels and tubes: A comprehensive review", *Alex. Eng. J.,* vol. 70, pp. 349-376, 2023. [http://dx.doi.org/10.1016/j.aej.2023.02.013]

[29] Q. Li, Y. Liu, S. Guo, and H. Zhou, "Solar energy storage in the rechargeable batteries", *Nano Today,* vol. 16, pp. 46-60, 2017. [http://dx.doi.org/10.1016/j.nantod.2017.08.007]

[30] R. Kumar, D. Lee, Ü. Ağbulut, S. Kumar, S. Thapa, A. Thakur, R.D. Jilte, C.A. Saleel, and S. Shaik, "Different energy storage techniques: recent advancements, applications, limitations, and efficient utilization of sustainable energy", *J. Therm. Anal. Calorim.,* vol. 149, no. 5, pp. 1895-1933, 2024. [http://dx.doi.org/10.1007/s10973-023-12831-9]

[31] T. Sood et al., "A review on thermal energy storage," in AIP Conference Proceedings, AIP Publishing, 2022. [http://dx.doi.org/10.1063/5.0080582]

[32] A. Thakur, R. Kumar, D. Lee, S. Kumar, P. Kumar, M. Asif, and Ü. Ağbulut, "Improved performance evaluation of a dual passage solar thermal collector using staggered turbulators", *Therm. Sci. Eng. Prog.,* vol. 53, p. 102668, 2024. [http://dx.doi.org/10.1016/j.tsep.2024.102668]

[33] A. Singhy, R. Thakur, R. Kumar, S. Kumar, S. Kumar, S. Kumar, and S. Thapa, "Influence of active water stream, irradiance, ambient temperature and wind speed on the efficiency of Fresnel lens based two stage PVT system", *Therm. Sci.,* vol. 26, suppl. 2 Part A, pp. 1139-1150, 2022. [http://dx.doi.org/10.2298/TSCI200801193S]

[34] A. Singhy, R. Thakur, and R. Kumar, "Experimental analysis for co-generation of heat and power with convex lens as SOE and linear Fresnel Lens as POE using active water stream", *Renew. Energy,* vol. 163, pp. 740-754, 2021. [http://dx.doi.org/10.1016/j.renene.2020.08.132]

[35] T. Sood, M.S. Al Ansari, T. Lakshmana Kishore, L. Malleswara Rao, J. Menaria, M.R. Mallu, D. Tyagi, K.S. Prasad, N.H. Vasoya, C. Balakrishna Moorthy, and P. Kumar, "A comprehensive review of the influence of nanomaterials on the thermal performance of a solar thermal collectors", *Mater. Today Proc.,* 2024. [http://dx.doi.org/10.1016/j.matpr.2024.05.010]

[36] Q. Alkhalaf, R. Kumar, A.R.S. Suri, S. Thapa, D. Lee, M. Alwetaishi, and Ü. Ağbulut, "A critical discussion of modelling, performance assessment, and design recommendations-based case study of solar dish Stirling system", *J. Therm. Anal. Calorim.,* vol. 150, no. 3, pp. 1999-2023, 2025. [http://dx.doi.org/10.1007/s10973-024-13879-x]

[37] C. Ahlem, B. A, D. i, and B. Barkati, "Comparative study of two control strategies proportional integral and fuzzy logic for the control of a doubly fed induction generator dedicated to a wind application", *International Journal of Power Electronics and Drive Systems (IJPEDS),* vol. 11, no. 1, p. 263, 2020. [http://dx.doi.org/10.11591/ijpeds.v11.i1.pp263-274]

[38] D.A. Taylor, *The design and testing of a horizontal axis wind turbine with sailfoil blades.* Open University: United Kingdom, 1986.

[39] D. Perry, *Aerodynamic design and structural analysis procedure for small horizontal-axis wind turbine rotor blade.* California Polytechnic State University, 2015. [http://dx.doi.org/10.15368/theses.2015.46]

[40] V. Quaschning, *Understanding renewable energy systems.* Routledge, 2014. [http://dx.doi.org/10.4324/9781315800493]

[41] V. Verma, *Highly Efficient Thermal Renewable Energy Systems: Design, Optimization and Applications.* CRC Press, 2024. [http://dx.doi.org/10.1201/9781003472629]

[42] R. Looney, *Handbook of oil politics.* Routledge, 2012. [http://dx.doi.org/10.4324/9780203851012]

[43] M.K. Farrell, *Lighting the Way into the Future: The Future of Energy Economics.* City University of New York Bernard Baruch College, 2024.

[44] H.C.O. Unegbu, D.S. Yawas, and B. Dan-asabe, "Assessing the environmental and economic benefits of solar energy integration in nigerian construction", *Preprint,* 2024. Version 1, available at Research Square [http://dx.doi.org/10.21203/rs.3.rs-XXXX/v1]

[45] P. Asmus, "Trends in the wind: Lessons from Europe and the U.S. in the development of wind power", *Corp. Environ. Strategy,* vol. 7, no. 1, pp. 51-61, 2000. [http://dx.doi.org/10.1016/S1066-7938(00)80114-1]

[46] W. Wang, G. Zheng, and J. Pan, *China's climate change policies.* Routledge Abingdon, 2012.

[47] S. Kumar, S. Sundaramurthy, D. Kumar, and A. K. Chandel, "Clean Energy Transition-via-Biomass Resource Utilization." [http://dx.doi.org/10.1007/978-981-97-6321-4]

[48] Y.T. Shah, *Modular systems for energy usage management.* CRC Press, 2020. [http://dx.doi.org/10.1201/9780367822392]

[49] A.J.M. Al Shahwani, *Synchromodal Container Transportation Towards Maritime Logistics Resilience: A Strategic Analysis in Pursuit of Sustainable Maritime Development.* Hamad Bin Khalifa University: Qatar, 2024.

[50] J. A. Stanturf, "Achieving sustainable management of boreal and temperate forests," 2020.

[51] Y.T. Shah, *Energy and fuel systems integration.* CRC Press, 2015. [http://dx.doi.org/10.1201/b19100]

[52] P.J. Gisi, *The Dark Factory and the future of manufacturing: A Guide to Operational Efficiency and Competitiveness.* CRC Press, 2024. [http://dx.doi.org/10.4324/9781032688152]

[53] F. Kreith, and D.Y. Goswami, *Energy management and conservation handbook.* CRC Press, 2007.

[54] T. Chanchayanon, S. Chaiprakaikeow, A. Jotisankasa, and S. Inazumi, "Optimization of Geothermal Heat Pump Systems for Sustainable Urban Development in Southeast Asia", *Smart Cities,* vol. 7, no. 3, pp. 1390-1413, 2024. [http://dx.doi.org/10.3390/smartcities7030058]

[55] R.M. Yonk, R.T. Simmons, and B.C. Steed, *Green vs. Green: the political, legal, and administrative pitfalls facing green energy production.* Routledge, 2013. [http://dx.doi.org/10.4324/9780203116135]

[56] J. Moreno, L. Campagnolo, B. Boitier, A. Nikas, K. Koasidis, A. Gambhir, M. Gonzalez-Eguino, S. Perdana, D-J. Van de Ven, A. Chiodi, E. Delpiazzo, H. Doukas, M. Gargiulo, A. Herbst, K. Al-Dabbas, Ş. Alibaş, F. Neuner, P. Le Mouël, and M. Vielle, "The impacts of decarbonization pathways on Sustainable Development Goals in the European Union", *Communications Earth & Environment,* vol. 5, no. 1, p. 136, 2024. [http://dx.doi.org/10.1038/s43247-024-01309-7]

[57] S. Chen, Q. Zhang, P. Andrews-Speed, and B. Mclellan, "Quantitative assessment of the environmental risks of geothermal energy: A review", *J. Environ. Manage.,* vol. 276, p. 111287, 2020. [http://dx.doi.org/10.1016/j.jenvman.2020.111287] [PMID: 32877889]

[58] W. Legrand, J.S. Chen, and G.C. Laeis, *Sustainability in the hospitality industry: Principles of sustainable operations.* Routledge, 2022.

[59] V. Minea, *Heating and Cooling with Ground-source Heat Pumps in Cold and Moderate Climates:*

*Design Principles, Potential Applications and Case Studies.* CRC Press, 2022.

[60] K.J. Skipka, and L. Theodore, *Energy resources: availability, management, and environmental impacts.* vol. Vol. 11. CRC Press, 2014. [http://dx.doi.org/10.1201/b16782]

[61] G. Axelsson, "The future of geothermal energy", In: *Living with Climate Change.* Elsevier, 2024, pp. 397-422. [http://dx.doi.org/10.1016/B978-0-443-18515-1.00009-5]

[62] D. Chandrasekharam, and J. Bundschuh, *Low-enthalpy geothermal resources for power generation.* CRC Press, 2008. [http://dx.doi.org/10.1201/9780203894552]

[63] T. Kashyap, R. Thakur, G.H. Ngo, D. Lee, G. Fekete, R. Kumar, and T. Singh, "Silt erosion and cavitation impact on hydraulic turbines performance: An in-depth analysis and preventative strategies", *Heliyon,* vol. 10, no. 8, p. e28998, 2024. [http://dx.doi.org/10.1016/j.heliyon.2024.e28998] [PMID: 38644884]

[64] S. Pokhrel, Computational Modeling of A Williams Cross Flow Turbine. 2017.

[65] K. Kashyap, R. Thakur, R. Kumar, and S. Kumar, "Feasibility analysis for conversion of existing traditional watermills in Western Himalayan region of India to micro-hydropower plants using a low head Archimedes screw turbine for rural electrification", *International Journal of Ambient Energy,* vol. 43, no. 1, pp. 7463-7473, 2022. [http://dx.doi.org/10.1080/01430750.2022.2068056]

[66] C. Kabre, *Synergistic Design of Sustainable Built Environments.* CRC Press, 2020. [http://dx.doi.org/10.1201/9781003102960]

[67] M.K. Singla, M. Safaraliev, J. Gupta, M. Aljaidi, I. Odinaev, R. Kumar, and A. Abdel Menaem, "Optimizing proton exchange membrane fuel cell parameter identification using enhanced hummingbird algorithm", *Int. J. Hydrogen Energy,* vol. 89, pp. 342-352, 2024. [http://dx.doi.org/10.1016/j.ijhydene.2024.09.211]

[68] S. Capareda, *Introduction to renewable energy conversions.* CRC Press, 2019. [http://dx.doi.org/10.1201/9780429199103]

[69] B.D. Fath, *Managing Air Quality and Energy Systems.* CRC Press, 2020. [http://dx.doi.org/10.1201/9781003043461]

[70] A. Kumar, Y. Liu, and R. Kumar, *Handbook of Intelligent and Sustainable Manufacturing: Tools, Principles, and Strategies.* CRC Press, 2024. [http://dx.doi.org/10.1201/9781003405870]

[71] A. Stuhlmacher, C.-W. Ten, L. Dilworth, and Y. Tang, "Operational planning for emerging distribution systems: A unique perspective on grid expansion", *Foundations and Trends® in Electric Energy Systems.,* vol. 7, no. 2, pp. 63-164, 2023. [http://dx.doi.org/10.1561/9781638283010]

[72] P. Rajca, A. Skibiński, A. Biniek-Poskart, and M. Zajemska, "Review of selected determinants affecting use of municipal waste for energy purposes", *Energies,* vol. 15, no. 23, p. 9057, 2022. [http://dx.doi.org/10.3390/en15239057]

[73] R. Belu, *Building Electrical Systems and Distribution Networks: An Introduction.* CRC Press, 2020. [http://dx.doi.org/10.1201/9780429173455]

[74] D.R. Patrick, S.W. Fardo, R.E. Richardson, and B.W. Fardo, *Energy conservation guidebook.* River Publishers, 2020. [http://dx.doi.org/10.1201/9781003151883]

[75] M. Asif, G. Sahin, and M. Khalid, *Handbook of Energy and Environment in the 21st Century: Technology and Policy Dynamics.* CRC Press, 2024.

[http://dx.doi.org/10.1201/9781032715438]

[76] D. Hidalgo, J.M. Martín-Marroquín, and F. Corona, "Biodegradable wastes in bioeconomy", In: *Waste Management in the Circular Economy.* Springer, 2024, pp. 55-76.

[77] F. Lesniewska, and K. Steenmans, *Circular economy and the law: Bringing justice into the frame.* Routledge, 2023.

[78] P. Kaur, "Biomass Energy for Sustainable Development."

[79] S.A. Aransiola, *Marine Greens: Environmental, Agricultural, Industrial and Biomedical Applications.* CRC Press, 2024.
[http://dx.doi.org/10.1201/9781003369738]

[80] L.M. Shaker, J.K. Mohammed, A. Basem, R.J. Halbos, R.R. Mahdi, S.A. Mohammed, M.A. Fayad, A. Al-Amiery, and M.H. Al Lami, "Comparative analysis of solar cells and hydrogen fuel: A mini-review", *Results in Engineering,* vol. 23, p. 102507, 2024.
[http://dx.doi.org/10.1016/j.rineng.2024.102507]

[81] S.M. Pinnoo, *Determining Geothermal Resource Potential of Northeast Oklahoma Through Machine Learning Techniques.* The University of Tulsa, 2023.

[82] S. Thapa, R. Kumar, K. Kumar, S. Samir, D. Lee, and T. Singh, "Analytical investigation of solar parabolic trough receiver with multiple twisted tapes using exergy and energy analysis", *Case Stud. Therm. Eng.,* vol. 50, p. 103493, 2023.
[http://dx.doi.org/10.1016/j.csite.2023.103493]

[83] Y.T. Shah, *Chemical energy from natural and synthetic gas.* CRC Press, 2017.
[http://dx.doi.org/10.1201/9781315302355]

[84] D.C. Makepa, and C.H. Chihobo, "Sustainable pathways for biomass production and utilization in carbon capture and storage—A review", *Biomass Convers. Biorefin.,* pp. 1-23, 2024.

[85] S. Thapa, R. Kumar, and D. Lee, "Energetic and exergetic analysis of jet impingement solar thermal collector featuring discrete multi-arc shaped ribs absorber surface", *Energy,* vol. 306, p. 132392, 2024.
[http://dx.doi.org/10.1016/j.energy.2024.132392]

[86] Y.T. Shah, *Thermal energy: Sources, recovery, and applications.* CRC Press, 2018.
[http://dx.doi.org/10.1201/9781315305950]

[87] A.K. Yadav, R. Kumar, M. Wang, G. Fekete, and T. Singh, "Comparative analysis of daily global solar radiation prediction using deep learning models inputted with stochastic variables", *Sci. Rep.,* vol. 15, no. 1, p. 10786, 2025.
[http://dx.doi.org/10.1038/s41598-025-95281-7] [PMID: 40155686]

[88] J. Bundschuh, and J. Hoinkis, *Renewable energy applications for freshwater production.* CRC Press, 2012.
[http://dx.doi.org/10.1201/b12521]

# CHAPTER 3

# Hydrogen Production from Biomass: Advancing Renewable Energy Technology

**Chetan Chauhan**[1,*]

[1] *Department of Chemistry, Sardar Patel University, Mandi 175001, Himachal Pradesh, India*

**Abstract:** $H_2$ production from biomass substrates offers a sustainable solution and carbon-neutral path to conventional hydrogen production methods. This chapter examines the potential role of thermochemical, biochemical, and biocatalytic processes for converting biomass into bio-hydrogen, evaluating their efficiency, scalability, and impact on the global clean energy transition challenges such as feedstock variability, process optimization, and economic feasibility are addressed to overcome such barriers. The chapter concludes with insights into future research, innovation, and policy support needed to advance biomass-based hydrogen production as a key renewable energy technology.

**Keywords:** Biomass energy, Hydrogen production, Renewable energy, Thermochemical processes.

## INTRODUCTION

Hydrogen is a futuristic and clean energy solution for addressing global energy security and environmental concerns with potential applications across the sectors, including electricity generation and transportation due to its suitability for transportation, long-term storage, and minimal environmental impact [1 - 3]. Among various contemporary fuels preferred for observed significant parameters (GF, EIF, and HCF), hydrogen gas has emerged as a primary clean and sustainable fuel (Table **1**). Major contemporary hydrogen production processes (about 95%) are fossil fuel-based and energy-intensive, including hydrocarbon reforming (Fig. **1**) [4]. Unlike most of the non biomass based contemporary $H_2$ production techniques, the biomass-based hydrogen production is considered carbon-neutral because it releases only $CO_2$ that was recently captured from the atmosphere during biomass growth, maintaining a balanced carbon cycle that

* **Corresponding author Chetan Chauhan**: Department of Chemistry, Sardar Patel University, Mandi 175001, Himachal Pradesh, India; E-mail: chetan@spumandi.ac.in

**Reetu Malhotra, Pooja Mahajan, Manish Singla & Ramesh Kumar (Eds.)**

distinguishes it from fossil-based hydrogen production, which introduces new carbon into the atmosphere and contributes to climate change *via* greenhouse gas emissions [5 - 7]. The current chapter explores recent advancements and challenges in the production domain of biomass-derived hydrogen production. Goren *et al.* highlighted the environmental and economic impacts of hydrogen production from contemporary sources, positioning biomass as a key player in sustainable hydrogen generation [8, 9]. On the global scenario, the hydrogen economy has been advancing into various methodologies for hydrogen generation, focusing on their performance and sustainability through innovation [10, 11].

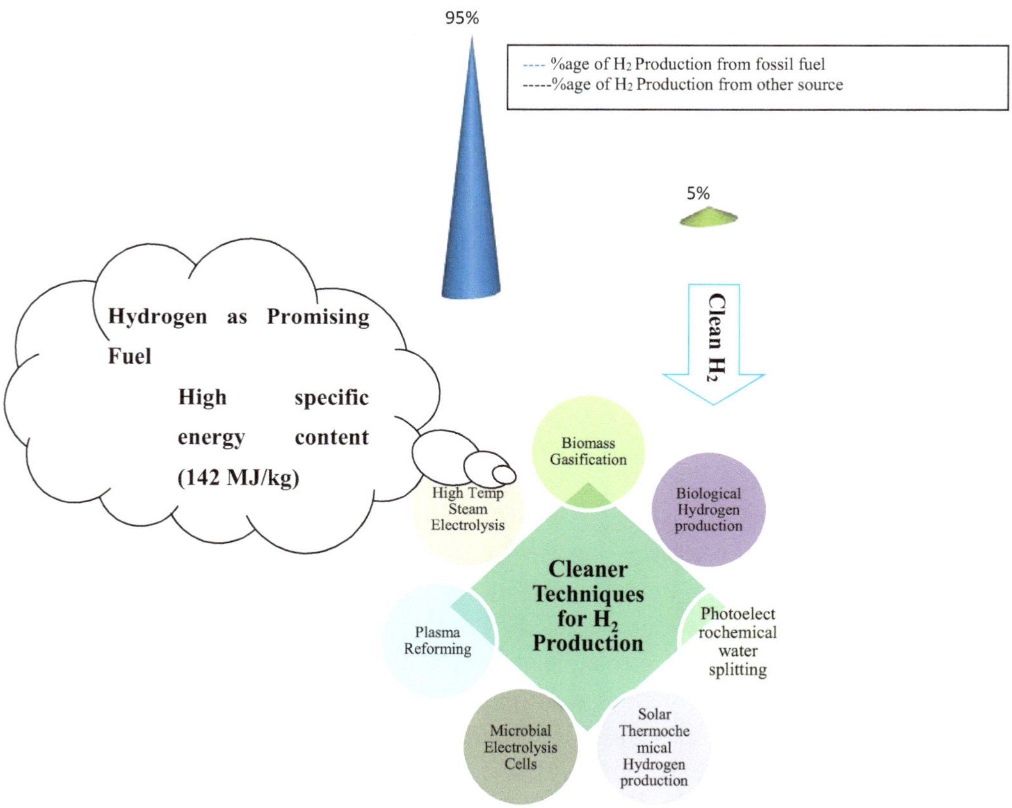


**Fig. (1).** Current status and pattern of H2 production in the global hydrogen economy based upon their sources.

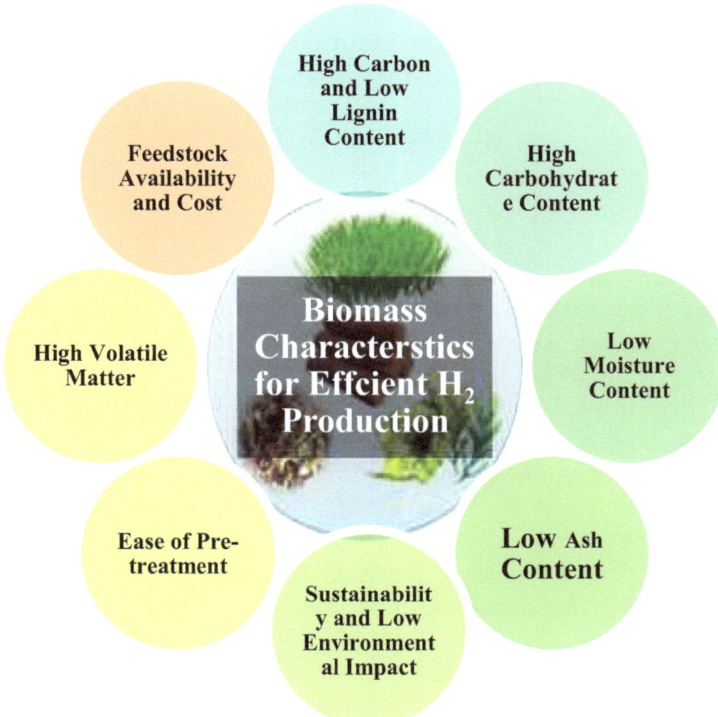


**Fig. (2).** Characteristics of biomass well-suited for biohydrogen production.

**Table 1. Comparison of contemporary energy fuels in terms of GF, EIF, and HCF.**

| Fuel Type | GF (Greenization factor) | EIF (Environmental impact factor) | HCF (Hydrogen content factor) |
|---|---|---|---|
| Hydrogen | 0.9 | 0.2 | 1.0 |
| Coal | 0.2 | 0.8 | 0.1 |
| Oil | 0.3 | 0.7 | 0.2 |
| Natural Gas | 0.4 | 0.6 | 0.3 |

Biomass is an organic resource that includes agricultural residues, forest residues, dedicated energy crops, organic municipal waste, and animal wastes, and can be explored as a sustainable solution for various applications [12, 13]. These organic matters can be explored to produce hydrogen and other byproducts through various methods, including gasification [14]. Variability in the characteristics of bio-feedstock is vital for yield and effective conversion into hydrogen production from the respective source (Fig. **2**). The following characteristics of biomass have an impact on efficient conversion and yields of $H_2$:

*High Carbon and Low Lignin Content*: Biomass with a higher carbon-to-lignin ratio is ideal, as lignin is resistant to microbial and thermochemical degradation, which can limit hydrogen production efficiency.

*High Carbohydrate Content*: Biomass rich in carbohydrates, especially cellulose and hemicellulose, provides an abundant source of fermentable sugars for biochemical hydrogen production processes.

*Low Moisture Content*: Biomatter with lower moisture is preferred for thermochemical processes like gasification and pyrolysis, as a higher percentage of water content can decrease the efficiency of the conversion process.

*Low Ash Content*: Biomass with lower ash %age reduces the risk of slagging and fouling in thermochemical processes, thus improving the system's performance and reducing maintenance requirements.

*Feedstock Availability and Cost*: Agricultural residues, forestry waste, and energy crops-based feedstock are the most easily available and abundant; therefore, such biomaterials are usually preferred for mass-scale hydrogen production [15].

*High Volatile Matter*: Biomass with a high percentage of volatile content is preferred for gasification, as it accelerates the yields of hydrogen production during thermal decomposition.

*Ease of Pre-treatment*: Biomass that can be easily pre-treated (mechanically, chemically, or biologically) to break down its complex structure allows for more efficient conversion into hydrogen.

*Sustainability and Low Environmental Impact*: Biomass should be sustainably sourced with minimal land-use competition and low associated greenhouse gas emissions to align with the environmental goals of hydrogen production.

These characteristics determine the suitability and efficiency of biomass in various biohydrogen productions. There are a number of approaches that have been used to explore the biomass for bio-hydrogen production. Each method of biohydrogen production from biomass has its strengths and contributes differently to the overall hydrogen yield. Biomass gasification and pyrolysis are the dominant methods due to their efficiency and versatility, while biological and hybrid methods offer complementary benefits and potential for future development (Fig. **3**). Understanding the contributions of each method helps in optimizing biohydrogen production strategies and advancing towards more sustainable energy solutions [1 - 15].

Such a hydrogen production process can further be classified broadly into the thermo-chemical and biochemical methods. These methods have their own salient features and provide an option for production (Table **2**).

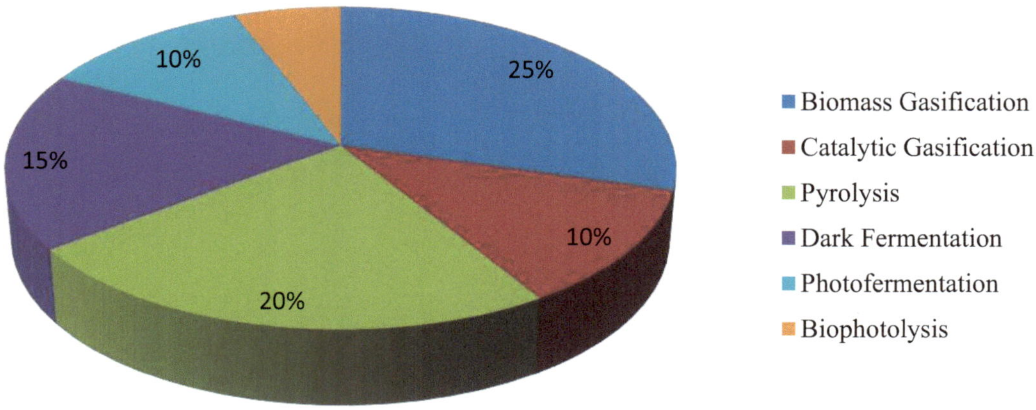


**Fig. (3).** Comparison and distribution of biohydrogen production by different methods.

**Table 2. Characteristics of thermo-chemical and biochemical methods.**

| Parameter | Thermo-chemical Methods[16–40] | Biochemical Methods[41–71] |
|---|---|---|
| Principle | Use heat and chemical reactions. | Use microorganisms and enzymes. |
| Common Processes | Pyrolysis, Gasification, Steam Reforming. | Dark Fermentation, Photofermentation, Biophotolysis |
| Temperature Requirement. | High (typically 700–1500 °C). | Low to moderate (ambient to ~60 °C). |
| Reaction Speed. | Fast. | Relatively slow. |
| Energy Source. | Thermal energy. | Solar/light energy and biological metabolism. |
| Feedstock. | Biomass, fossil fuels, and natural gas. | Organic waste, carbohydrates, and organic acids. |
| Catalyst/Agent. | Chemical catalysts (e.g., Ni for reforming). | Microbial enzymes, bacteria, and algae. |
| Hydrogen Yield. | High. | Moderate to low. |
| Environmental Impact. | May generate $CO/CO_2$ depending on the process. | Low emissions; more environmentally benign. |
| Process Complexity. | Complex systems require high-temperature reactors. | Simpler setups but sensitive to biological conditions. |
| Scalability. | High. | Limited by biological constraints. |
| Example Output. | $H_2$-rich syngas ($H_2$ + CO), purified hydrogen. | $H_2$ with $CO_2$ and volatile organic byproducts. |

## THERMO-CHEMICAL PROCESSES FOR $H_2$PRODUCTION

Thermochemical conversion is one class of the most explored methods for hydrogen production, where biomaterial is subjected to higher temperatures [16 - 18]. Thermochemical methods are highly efficient for converting a variety of biomass resources into hydrogen-rich syngas. Advancements in thermochemical processes with effective reactor designs and catalysts have greatly improved hydrogen yields and energy efficiency [17 - 23]. Further, the integration of thermochemical with water splitting has demonstrated significant improvements in hydrogen generation efficiency [20].

### Biomass Gasification

Gasification is a promising thermos-chemical based technology for mass-scale production of hydrogen [24]. Gasification involves the partial combustion of biomatter in a controlled environment to produce syngas, subsequently followed by critical cleanup steps for purity check just after the hydrogen separation [25]. Gasification technology has advanced significantly, especially in terms of integrating biomass feedstocks with improved gasification reactors [22]. Biomass gasification has been influenced by various factors as listed below in Fig. (**4**).

The variability in the bio-feedstock characteristics plays a crucial role in the performance of the gasification process [26]. High moisture content reduces gasification efficiency, as more energy is required to evaporate the water, lowering the temperature and hydrogen yield. For efficient gasification, the biomass moisture content should ideally be below 20%. Smaller biomass particles increase the surface area available for reaction, enhancing the conversion rate. However, very fine particles can lead to operational issues like agglomeration in fluidized bed reactors. High ash content in biomass can lead to slagging and fouling in reactors. Ash can also act as a heat sink, reducing the temperature and slowing down reactions. The proportion of cellulose, hemicellulose, and lignin affects the gasification process. Lignin, being more resistant to thermal degradation, requires higher temperatures for gasification. In biomass gasification, temperature is one of the key factors in determining the efficiency, product composition, and distribution. In low-temperature gasification (500°C to 700°C), the process requires less energy, which lowers operational costs. Nickel-based or AAEM catalysts effectively enhance hydrogen production in better control over reaction by reducing and promoting the WGS (Water-gas Shift reaction)and tar formation [27]. However, low temperatures can lead to higher char formation and are less effective at breaking down complex tars without catalysts. In High Temperature Gasification (800°C to 1000°C or above), the process achieves higher conversion efficiency, effectively breaking down biomass into simpler gaseous products like hydrogen, carbon monoxide, and methane with minimal tar formation [24]. Higher temperatures also improve reaction rates and reduce char formation. However, the increased energy input raises operational costs, and catalysts can deactivate due to sintering, leading to additional wear and tear on reactor materials. The choice of gasifying agents like air, oxygen, steam, or a mixture directly impacts the syngas composition and hydrogen yield [28]. The use of air as a gasifying agent is cost-effective but leads to nitrogen dilution in the syngas, resulting in a lower hydrogen concentration. Pure oxygen improves hydrogen yield by eliminating nitrogen dilution, but it increases operational costs due to oxygen production. Steam gasification (with or without oxygen) enhances the efficiency of the Water-gas Shift (WGS) reaction, increasing the percentage of hydrogen in the syngas. Sometimes $CO_2$ is used to enhance reactions like dry reforming, which increases $H_2$ production by reacting with hydrocarbons in the gas phase. Pressure: Operating at high pressure (typically between 10 and 30 bar) can increase the hydrogen yield by favoring certain reactions like steam reforming. It also helps in easier syngas compression for downstream processes like hydrogen separation. Atmospheric pressure gasification is simpler and less costly but may result in lower hydrogen yields compared to pressurized systems.

The time the biomass spends in the reactor, or residence time, is another important factor. Longer residence times ensure more complete gasification of biomass, leading to a higher hydrogen yield. However, if the residence time is too short, unreacted biomass can result in low gas yields. Different reactor designs influence gasification efficiency, including entrained-flow, plasma reactors, fixed-bed, and fluidized-bed [29]. Fixed-bed reactors are simpler but tend to produce more tar and have lower hydrogen yields. Fluidized-bed Reactors offer better heat and mass transfer, leading to higher hydrogen production and more uniform temperature distribution [30]. High temperatures and pressures are the preferred operation conditions for entrained flow reactors for promoting high hydrogen yields and better conversion of biomass with minimal tar formation [31].

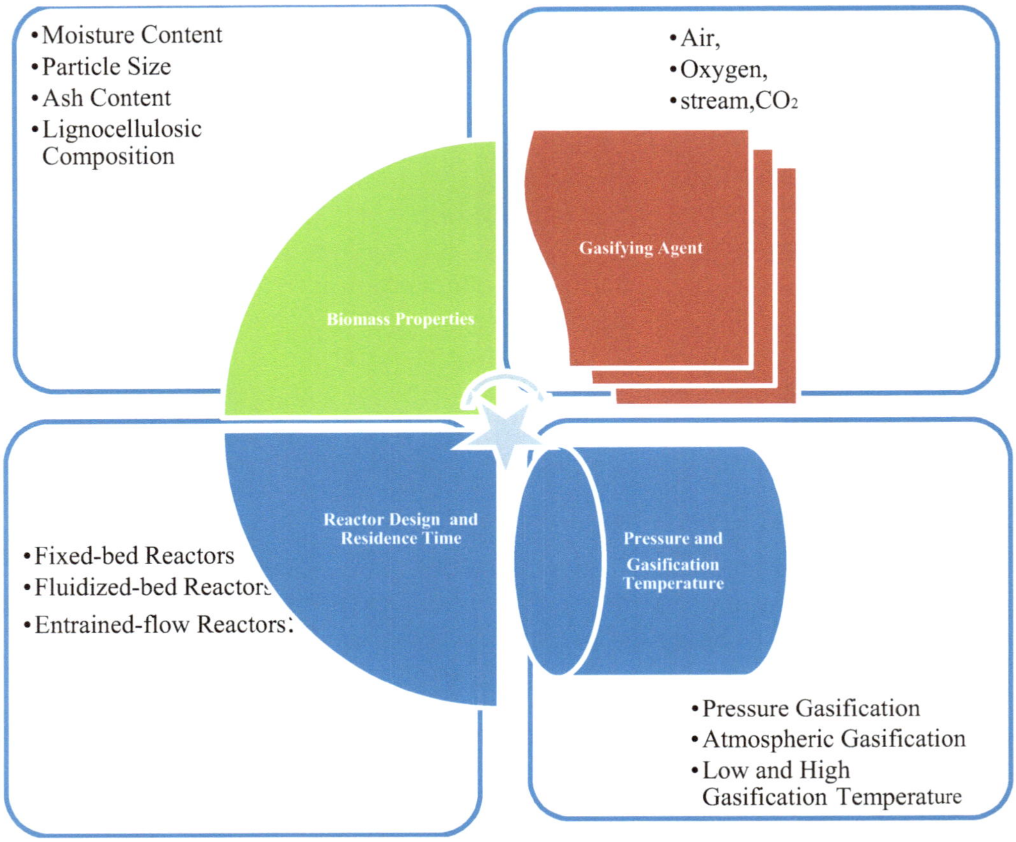


**Fig. (4).** Factors influencing Biomass gasification.

## Catalytic Gasification

Recent advances in catalysis have demonstrated the vital role in the improvement of gasification efficiency by promoting the breakdown of hydrocarbons through the application of advanced catalysts [13] [32]. Various improved catalytic cleaning methods for syngas production have been demonstrated in the gasification process [24]. Faizan and Song highlighted key challenges that include the deactivation and the higher cost of catalysts, along with the most plausible mitigation steps [33]. The key steps involved in the catalytic gasification from biomass are illustrated in Table **3** and Fig. (**5**).

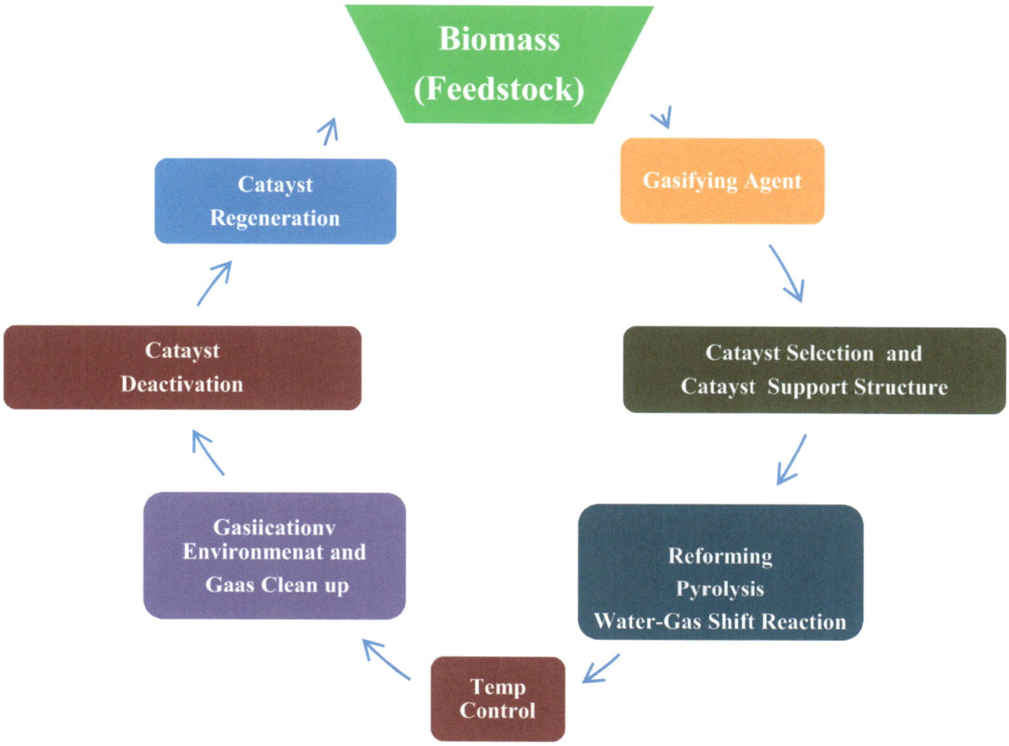


**Fig. (5).** Key steps involved in the catalytic gasification of biomass.

**Table 3. Description of key aspects of catalytic gasification from biomass [24, 31 - 35].**

| Step | Description |
|---|---|
| Catalyst Selection (based on desired benefits ) | Nickel-Based Catalysts: Effective in reforming hydrocarbons and steam gasification; prone to sintering and carbon deposition. AAEM Catalysts: Promote biomass conversion, reduce tar formation, and increase $H_2$ yield *via* (WGS) Water-gas Shift reaction.- Iron-Based Catalysts: Effective in tar cracking, can handle high-sulfur biomass feedstocks. Zeolites: Used for gas conditioning and improving hydrogen purity. |
| -Catalyst Support and Structure | Support Materials: Catalysts are supported on materials like alumina, silica, or magnesium oxide to enhance surface area and stability.- Structure: Structured catalysts (monoliths, foams) enhance mass transfer, reduce pressure drop, and improve contact between gas and catalyst, increasing hydrogen yields. |
| Temperature Control | Lower Temperatures (500°C-700°C): Catalysts allow gasification at lower temperatures, improving energy efficiency and reducing tar formation.- Higher Temperatures (>800°C): While typically required for non-catalytic gasification, catalysts reduce the temperature needed, enhancing energy efficiency in hydrogen production. |

*(Table 3) cont.....*

| Step | Description |
|---|---|
| Managing Catalyst Deactivation | Sintering: High temperatures cause agglomeration of catalyst particles, reducing surface area.- Carbon Deposition: Carbon builds up on catalysts, reducing active sites; steam reforming helps gasify the carbon to restore efficiency.- Sulfur Poisoning: Sulfur compounds in biomass can poison catalysts, especially nickel-based ones. Pre-treatment or the use of sulfur-tolerant catalysts is required. |
| Gasification Environment | Steam-to-Biomass Ratio: Affects hydrogen production; higher steam content promotes the water-gas shift reaction, but excessive steam reduces energy efficiency.- Pressure: Operating at elevated pressure favors hydrogen production but increases operational complexity and cost. |
| Catalyst Regeneration | -Catalysts undergo periodic regeneration (*via* air or steam) to remove coke deposits and restore catalytic performance. |

## Pyrolysis

In the pyrolysis, the thermal decomposition of a biomatter occurs in the absence of oxygen to produce bio-oil, syngas, and biochar. $H_2$production can be achieved, from the catalytic steam reforming of bio-oil [36]. Pangestu *et al.* explored the potential of methane pyrolysis as an alternative method for sustainable hydrogen production, whereas Setiabudi *et al.* focused on the use of catalytic steam reforming for producing hydrogen from biomass-derived oils [37 - 39]. Pyrolysis-based technology is promising, but further research is required to optimize the process catalyst performance and conditions, including temperature, heating rate, biomass composition, residence time, and reactor design [39, 40].

# BIOCHEMICAL METHODS FOR HYDROGEN PRODUCTION

Biochemical methods, including dark fermentation and photofermentation, offer an environmentally friendly pathway to hydrogen production using microorganisms to break down biomass (Table **4**). Biological hydrogen production processes emphasize that these methods are highly sustainable and can utilize waste biomass streams [41]. The potential of different microbial processes varies for hydrogen production, highlighting the importance of optimizing process parameters to increase hydrogen yields [41 - 43].

## Dark Fermentation

Dark fermentation is another promising technology where bacteria produce hydrogen in anaerobic conditions [40] [45, 46]. Unlike photofermentation, dark fermentation does not rely on sunlight, making it more flexible and easier to scale for continuous hydrogen production. The process primarily involves fermentative pathways that convert organic compounds into $H_2$, $CO_2$, and Volatile Fatty Acids

(VFAs) and alcohols as side product (Equation 1) [46].

**Table 4. Approaches for bio-hydrogen production by exploring biomass and photosynthetic microorganisms.**

| Approaches | | | Advantages | Limitations | References |
|---|---|---|---|---|---|
| Thermo-chemical Process | Gasification | Biomass Gasification | • Higher yield of $H_2$ production can process a wide variety of biomass types.<br>• Suitable for large-scale production.<br>• Carbon-neutral if managed sustainably.<br>• Can utilize waste biomass.<br>• Well-established technology. | • Requires high temperature and energy input.<br>• Costly gas cleanup processes are used to remove impurities.<br>• Syngas produced contains impurities.<br>• Tar removal is challenging and adds cost. | [24-31] |
| | | Catalytic Gasification | • Improved efficiency and hydrogen yield.<br>• Reduction in tar formation with the right catalyst. | • Catalyst deactivation over time.<br>• High operational and catalyst regeneration costs. | [24],[31-35] |
| | Pyrolyss | | • Can produce hydrogen-rich syngas.<br>• Effective for processing lignocellulosic biomass. | • Requires high temperatures.<br>• Bio-oil from pyrolysis requires further processing for hydrogen production. | [36-40] |

*(Table 4) cont.....*

| Approaches | | | Advantages | Limitations | References |
|---|---|---|---|---|---|
| Biochemical and Microbial Electrochemical Methods | Fermentation | Dark Fermentation | • Operates under mild conditions. • Can use organic waste streams. | • Low hydrogen yield. • By-products (e.g., organic acids) need to be managed. | [13],[26] [40-48] |
| | | Photo Fermentation | • Renewable hydrogen production using sunlight. • Can utilize a wide range of biomass sources. | • Requires specialized light systems. • Low energy conversion efficiency. | [43-57] |
| | Electrochemical | Microbial Electrochemical (MEC) | • Sustainable and can be integrated with wastewater treatment. • Potential for continuous hydrogen production. | • Low hydrogen production rates. • High capital costs and system complexity. | [58-69] |
| Photolysis Methods | Direct | | • Evolve hydrogen directly from water by directly harvesting light | • Need for high-intensity light; lower photochemical efficiency; oxygen content inhibits the process | [51],[70-71] |
| | Indirect | | • Blue-green algae produce hydrogen and fix atmospheric nitrogen ($N_2$) | • Uptake hydrogenases must be removed; oxygen inhibits nitrogenase enzyme activity | |

$$C_6H_{12}O_6 \rightarrow 2CH_3CH_2OH + 2CO_2 + 2H_2 \quad \textbf{(1)}$$

To improve hydrogen production, some of the common reactor configurations are used in dark fermentation, including CSTRs (Continuous Stirred Tank Reactors), UASB (Up Flow Anaerobic Sludge Blanket Reactors), and fixed-bed reactors that emphasize maximizing the contact between microbes and substrates while allowing for the continuous removal of hydrogen and by-products. The *Clostridium* is the most commonly used bacterium, besides several other species, such as *Enterobacter* and *Thermoanaerobacterium,* that are also capable of producing hydrogen. Guo *et al.* (2010) investigated the $H_2$production with the key challenge of substrate limitations that need to be address from agricultural waste using dark fermentation [44].

Several key factors influence the efficiency and hydrogen yield of dark fermentation [40 - 46]:

*pH:* Maintaining an optimal pH value (5.5- 6.5 pH) is essential for maximizing hydrogen production and microbial activity. Too low or too high pH levels can inhibit hydrogen-producing bacteria and reduce the overall yield [13].

*Temperature:* Hydrogen production in dark fermentation is highly temperature-sensitive. Most of the mesophilic reaction conditions (30-40°C) are explored, although thermophilic bacteria can operate at 50-60°C to enhance hydrogen yield and reduce substrate degradation time [47].

*Substrate* Type: The selection of a substrate plays a vital role in hydrogen production. Simple sugars like glucose tend to produce higher hydrogen yields, but the use of complex biomass such as lignocellulosic materials or waste can reduce reaction rates. Pre-treatment of such cases of substrates can improve their bioavailability for bacteria [26].

*HRT (Hydraulic Retention Time):* Optimizing HRT is essential to balance hydrogen production with the microbial breakdown of the substrate. Longer retention times can increase hydrogen yield but reduce process efficiency.

*Bacterial Strains*: Different microbial strains or mixed microbial communities can influence the performance of dark fermentation. The microbial community is optimized through strain selection or genetic engineering to enhance $H_2$ production [48].

**Photofermentation**

Photofermentation is another sustainable approach from biological systems where the photosynthetic bacteria harvest the light energy and subsequently convert organic mass into $H_2$ gas *via* the nitrogenase enzyme (Fig. **6**) [49 - 53]. Photofermentation is primarily driven by anaerobic photosynthetic bacteria, including *Rhodobacter sphaeroides* and *Rhodopseudomonas palustris*, which are capable of producing hydrogen [51 - 56]. These bacteria harvest light as the primary energy source, and organic compounds (such as organic acids, sugars, or waste products from agriculture and industry) act as electron donors. The reaction for hydrogen production in photofermentation is shown in equation 2.

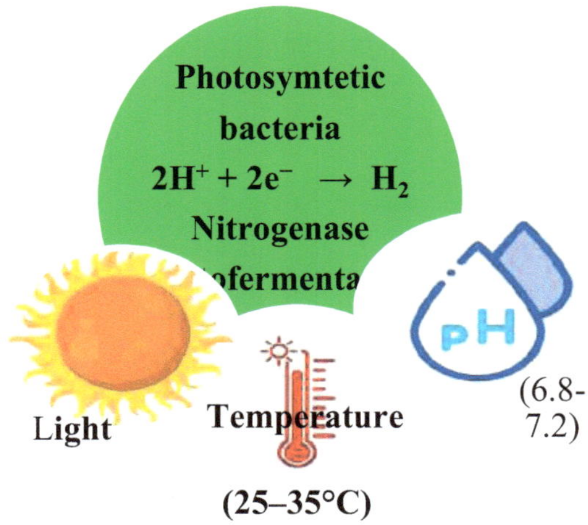


**Fig. (6).** Optimized Conditions for photofermentation.

$$\text{Organic Substrate} \rightarrow \text{Hydrogen } (H_2) + CO_2 + \text{Other Byproducts} \quad \textbf{(2)}$$

Unlike some biofuel production processes, photofermentation does not directly compete with food crops because it can use non-food organic substrates. The overall process can be divided into two key phases [51 - 57]:

### *Light Absorption and Electron Transport*

The bacteria absorb light through pigments like bacteriochlorophylls. The harvested light energy is transformed into chemical energy, driving the electron transport chain in the bacterial membrane. The light-driven reaction creates a proton gradient, which leads to the production of ATP [58].

### *Hydrogen Production*

Simultaneously, the bacteria metabolize organic matter through a series of reactions that release the protons and electrons and are subsequently utilized by a specific enzyme known as nitrogenase, to catalyze the further production of ($H_2$) in the absence of nitrogen. Nitrogenase requires a significant amount of ATP to function, and the light reactions of photosynthesis provide the necessary energy for this process. To maximize hydrogen production in photofermentation, specially designed bioreactors are used. These reactors are typically transparent to allow maximum light penetration and are designed to optimize the exposure of the bacterial culture to light. Several types of reactors, including flat-plate, tubular, and photobioreactors, are being investigated to enhance the efficiency of the process [51 - 57].

## Microbial Electrochemical Systems

Microbial Electrolysis Cells (MECs) represent a futuristic electrochemical technology that leverages electroactive microorganisms to catalyze the electrolysis of water or organic mass, to produce sustainable $H_2$ gas or other valuable chemicals [58 - 66]. By utilizing a variety of electrochemically active bacteria, organic matter, and wastewater, MECs offer a promising alternative solution to traditional methods. However, there are significant technological and economic challenges that need to be addressed for widespread adoption of such technology [67]. These systems stand out for their low energy consumption, self-sustaining microbial biocatalysts, high efficiency, and the ability to produce value-added products. Despite their potential, several challenges include low hydrogen production rates, high internal resistance, and the use of costly materials, which must be addressed to optimize the overall performance of MECs. There are several critical factors that play a vital role in enhancing MEC's efficiency that includes microbial species selection, electrode materials, substrate composition, and system design. Each of the MEC configurations offers distinct advantages and limitations depending on factors such as the specific application, performance goals, and cost effectiveness [17]. Researchers are actively investigating innovative designs and operational strategies to improve the efficiency and scalability of MEC technology for sustainable hydrogen production and other bio-electrochemical processes [68]. Advancements in areas such as biofilm engineering, electrode development, and reactor design will be essential for broader commercial applications. Therefore, Microbial Electrochemical Systems (MESs) offer a promising solution for sustainable energy production.

The Electrochemically Active Bacteria (EAB) are instrumental in the facilitation of electron transfer from biomass to electrodes in Microbial Electrolysis Cells (MECs). Such a mechanism is known as Extracellular Electron Transfer (EET), which involves transferring electrons to extracellular acceptors without an external mediator. EET occurs *via* direct electron transfer through outer membrane proteins, electron shuttles, or through pili acting as nanowires within the biofilm. MECs' performance is highly dependent on a number of factors, including microbial activity, electrode characteristics, membrane characteristics, reactor configuration, biomass quality, and operational conditions, including temperature, pH, and salinity [69]. Besides the above techniques, biophotolysis is another class of promising technology where light energy is harvested by photosynthetic microorganisms to be utilized in a water split reaction ($H_2O$) and subsequently converted into hydrogen ($H_2$) and oxygen ($O_2$) [70] [51]. Such a biological process is based on non-biomass feedstock and facilitated by microalgae, cyanobacteria, and other photosynthetic organisms that possess key enzymes and photosystems for light harvesting and water splitting with any harmful emissions [71]. Biophotolysis can be direct or indirect, depending on its mechanisms and pathways for hydrogen production.

## CASE STUDIES

There are several case studies that highlight the diverse pathways and potential of both thermo-chemical and biochemical approaches in advancing sustainable hydrogen production. Some case studies are illustrated in Table **5** [72, 73].

**Table 5. Comparative aspects of the efficiency metric and parameters for credibility.**

| **Case study** | **Thermo-chemical Methods[72]** | **Biochemical Methods [73]** |
|---|---|---|
| | **• GoBiGas Project, Gothenburg, Sweden** | **• *Clostridium butyricum*-based hydrogen production in lab fermenters** |
| Process type | • Biomass Gasification | • Dark Fermentation |
| Operation parameters/efficiency metric | | |
| Feedstock Used | • Forest biomass, wood chips | • Glucose, food waste, organic wastewater. |
| Operating Conditions | • High temperature (700–1500 °C), partial oxidation | • Mesophilic or thermophilic (30–60 °C), anaerobic. |
| Overall Efficiency | • Cold gas efficiency approx. 65% | • Conversion efficiency is typically 15–30%. |
| Hydrogen Yield | • 55% hydrogen, recoverable *via* water-gas shift | • Up to 2.5 mol hydrogen /mol glucose. |
| Scalability | • Demonstrated high performance at pilot and semi-commercial scale | • Mostly lab or pilot-scale studies. |
| By-products | • Ash, tar, carbon monoxide, carbon dioxide, and methane. | • Volatile fatty acids, carbon dioxide organic residues. |
| Water Usage | • Moderate, and used primarily for steam and quenching. | • Minimal water input may be for pretreatment. |
| Environmental Impact | • Depends on emissions control and feedstock. | • Lower emissions require effluent management. |
| Challenges | • Feedstock variability, material corrosion, and Tar formation. | • Lower hydrogen yield, accumulation of inhibiting microbes |
| Advancements | • Dual-fluidized beds and tar reforming, steam/oxygen optimization process. | • Integration of bio-reactors and genetic engineering, and co-culture use. |

## CHALLENGES AND SOLUTIONS

The global Sustainable Development Goals (SDGs) depend upon the successful integration of bio-hydrogen production into existing energy infrastructure. The challenges of hydrogen production from biomass are not only technical but also involve economic and policy considerations [71] [74 - 83]. Biomass-based hydrogen production faces several challenges that need to be addressed

for broader adoption. Adapting infrastructure for biomass-derived hydrogen involves modifications to storage and distribution systems. Scaling up for practical production requires optimizing processes, engineering innovations, resource management, safety protocols, market readiness, and continuous research (Fig. 7). Effectively addressing these challenges can position the hydrogen production techniques as a sustainable and viable solution, shifting to low-carbon energy in the contemporary era [71] [74 - 77].

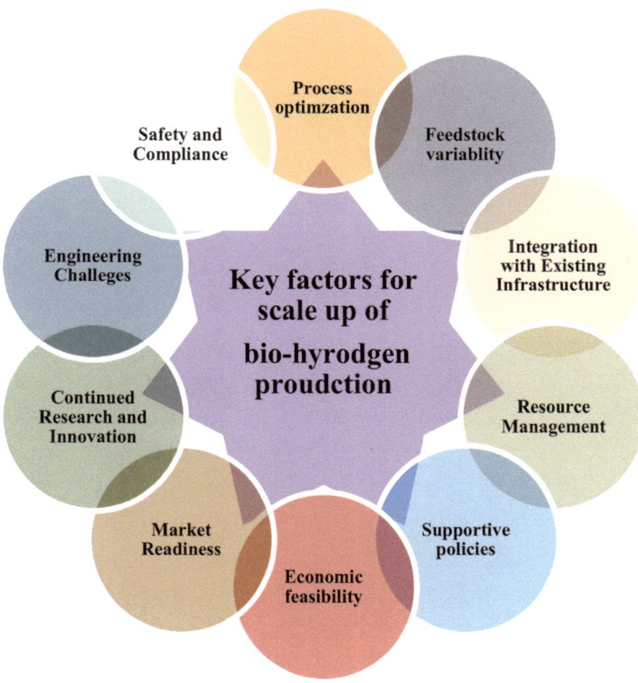


**Fig. (7).** Key factors for scale-up of bio-hydrogen production.

Variability in biomass feedstock can impact process efficiency. Solutions include developing robust pre-treatment and standardization techniques. Continued advancements in catalysts and microbial strains are essential to optimize hydrogen production and minimize by-products. Balancing economic feasibility with environmental benefits is critical. Integrating multiple methods, such as combining dark fermentation and photofermentation or coupling biological processes with thermochemical ones, can improve hydrogen yields and economic feasibility. Innovative integration approaches can improve atom economy and lead to higher hydrogen production efficiency. Advancements in genetic engineering upgrade the modification process of microorganisms to enhance their hydrogen-generating abilities by overcoming limitations such as low yield, oxygen sensitivity, and substrate specificity.

## FUTURE DIRECTIONS AND POLICY SUPPORT

The future of biomass-based hydrogen production depends on technological innovation through continuous research, supportive and effective policies for wider adoption and advancement of technology across the application domain [78 - 82]. Future research is a crucial step in advancing biomass-based hydrogen production technology that should focus on developing advanced thermochemical and biochemical methods, including MECs configuration, photo fermentation, *etc* [82 - 85]. To make green hydrogen a viable and cost-efficient clean energy source, there is an urgent need to pump huge investment in research and development, coupled with supportive policy frameworks. Without accelerating innovation, current hydrogen technologies will continue to struggle with high setup costs, limited scalability, and inefficient energy conversion [86]. Key improvements are needed in areas like catalytic materials, which can greatly enhance the performance of electrolysis and biomass-based hydrogen production. Additionally, advancements in the reactor and electrolyzer designs that will allow for flexible, modular systems could make deployment more efficient and adaptable to different scales [86]. Another promising approach lies in the integration of biorefinery systems. Such systems combine various biomass conversion methodologies, making it possible to produce hydrogen along with other valuable products while maximizing the atom economy and thus enhancing the overall economic feasibility [86]. However, feasibility for large-scale hydrogen production *via* such innovations will require strategic funding, strong collaborations among public and private players, and international cooperation. For hydrogen technologies to become a key part of a sustainable global energy future, it is crucial that governments, industries, and research institutions work together. This includes creating consistent regulatory policies, building necessary infrastructure, and investing in workforce development and training. A coordinated, long-term strategy is essential to overcome current barriers and unlock the full potential of hydrogen as a clean energy solution [86 - 89].

## CONCLUSION

Trends and advancements in biomass-based $H_2$ production technology demonstrate a futuristic solution towards economically viable, sustainable energy to attain SDGs and meet the energy security across the global energy sector. Advances in thermochemical and biochemical processes, through ongoing research, are paving the way for designing more efficient and scalable hydrogen production approaches. Addressing the upcoming technical and infrastructure challenges and leveraging effective policy support will be vital in realizing the full potential of a carbon-neutral solution from biomass-based hydrogen technology in the global growing energy needs and energy transition. Recent and key technological advancements in catalytic materials, modular reactor designs, and integrated biorefinery systems are needed to improve efficiency and flexibility in both thermochemical and biochemical processes. Strategic collaboration in research and development among governments, industries, and academia, along with supportive policy, is vital to overcome current barriers and challenges that remain, particularly in terms of scalability, cost, and integration with existing energy infrastructure, to unlock the full potential of biomass-derived hydrogen clean energy.

## REFERENCES

[1] K. Dali, and N. Pampoukidou, *"Recent advances in hydrogen production & Storage"*, University of Thessaly, 2022. https://ir.lib.uth.gr/xmlui/bitstream/handle/11615/59648/25177.pdf?sequence=1 Diploma Thesis

[2] B. Liu, L. Feng, H. Ayed, A. Mouldi, and R. Marzouki, "The dynamic modeling and the exergy assessment of hydrogen synthesize and storage system with the power-to-gas concept for various locations", *Int. J. Hydrogen Energy,* vol. 47, no. 62, pp. 26000-26012, 2022. [http://dx.doi.org/10.1016/j.ijhydene.2022.03.187]

[3] *New Technology to Produce Hydrogen from Renewable Energy Sources Based on AI with Optimized Costs for Environmental Applications..* Available from: https://www.era-learn.eu/network-information/networks/enerdigit/joint-call-2020-micall20-on-digital-transformation-for-green-energy-transition/new-technology-to-produce-hydrogen-from-renewable-energy-sources-based-on-ai-with-optimized-costs-for-environmental-applications

[4] J. Rey, F. Segura, and J. Andújar, "Green Hydrogen: Resources Consumption, Technological Maturity, and Regulatory Framework", *Energies,* vol. 16, no. 17, p. 6222, 2023. [http://dx.doi.org/10.3390/en16176222]

[5] S. Arora, M. Aljaidi, M. Sandhu, S.H.A. Khan, and R. Malhotra, "Determination of yearly energy savings by CVR in combination with photovoltaic inverter control functions in distribution networks", *Energy Reports,* vol. 12, pp. 5507-5520, 2024. [http://dx.doi.org/10.1016/j.egyr.2024.11.001]

[6] C. Acar, and I. Dincer, "Hydrogen Production", *Comprehensive Energy Systems,* vol. 3, pp. 1-40, 2018.

[7] J.R.C. Rey, C. Mateos-Pedrero, A. Longo, B. Rijo, P. Brito, P. Ferreira, and C. Nobre, "Renewable hydrogen from biomass: technological pathways and economic perspectives", *Energies,* vol. 17, no. 14, p. 3530, 2024. [http://dx.doi.org/10.3390/en17143530]

[8] A. Yagmur Goren, I. Dincer, and A. Khalvati, "A comprehensive review on environmental and economic impacts of hydrogen production from traditional and cleaner resources", *J. Environ. Chem. Eng.,* vol. 11, no. 6, p. 111187, 2023. [http://dx.doi.org/10.1016/j.jece.2023.111187]

[9] M.K. Singla, M. Safaraliev, J. Gupta, M. Aljaidi, I. Odinaev, R. Kumar, and A. Abdel Menaem, "Optimizing proton exchange membrane fuel cell parameter identification using enhanced hummingbird algorithm", *Int. J. Hydrogen Energy,* vol. 89, pp. 342-352, 2024. [http://dx.doi.org/10.1016/j.ijhydene.2024.09.211]

[10] S.O. Jeje, T. Marazani, J.O. Obiko, and M.B. Shongwe, "Advancing the hydrogen production economy: A comprehensive review of technologies, sustainability, and future prospects", *Int. J. Hydrogen Energy,* vol. 78, pp. 642-661, 2024. [http://dx.doi.org/10.1016/j.ijhydene.2024.06.344]

[11] K. Chelvam, M.M. Hanafiah, K.S. Woon, and K.A. Ali, "A review on the environmental performance of various hydrogen production technologies: An approach towards hydrogen economy", *Energy Rep.,* vol. 11, pp. 369-383, 2024. [http://dx.doi.org/10.1016/j.egyr.2023.11.060]

[12] Sources and technologies for green hydrogen production: A review. Available from: https://www.uwasa.fi/sites/default/files/2023-08/T1.1a%20Report%2007082023_final.pdf

[13] B. Correia, H. A. Matos, T. F. Lopes, S. Marques, and F. Gírio, "Sustainability assessment of 2G bioethanol production from residual lignocellulosic biomass", *Processes,* vol. 12, no. 5, p. 987, 2024. [http://dx.doi.org/10.3390/pr12050987]

[14] Hydrogen Production: Biomass Gasification, U.S. Department of Energy. Available from: https://www.energy.gov/eere/fuelcells/hydrogenproduction-biomass-gasification

[15] Second Generation Biofuels Market. Available from: https://www.verifiedmarketresearch.com/product/secondgeneration-biofuels-market/

[16] K. Bélafi-Bakó, G. Tóth, P. Bakonyi, and N. Nemestóthy, "Utilization of agro-wastes in biohydrogen fermentation by various microorganisms", *Hung. J. Ind. Chem.,* vol. 50, no. 2, pp. 57-60, 2022. [http://dx.doi.org/10.33927/hjic-2022-19]

[17] S.K. Dash, S. Chakraborty, and D. Elangovan, "A Brief Review of Hydrogen Production Methods and Their Challenges", *Energies,* vol. 16, no. 3, p. 1141, 2023. [http://dx.doi.org/10.3390/en16031141]

[18] G. Ji, J.G. Yao, P.T. Clough, J.C.D. da Costa, E.J. Anthony, P.S. Fennell, W. Wang, and M. Zhao, "Enhanced hydrogen production from thermochemical processes", *Energy Environ. Sci.,* vol. 11, no. 10, pp. 2647-2672, 2018. [http://dx.doi.org/10.1039/C8EE01393D]

[19] J.E. Lee, I. Shafiq, M. Hussain, S.S. Lam, G.H. Rhee, and Y.K. Park, "A review on integrated thermochemical hydrogen production from water", *Int. J. Hydrogen Energy,* vol. 47, no. 7, pp. 4346-4356, 2022. [http://dx.doi.org/10.1016/j.ijhydene.2021.11.065]

[20] M.A. Rosen, "Advances in hydrogen production by thermochemical water decomposition: A review", *Energy,* vol. 35, no. 2, pp. 1068-1076, 2010. [http://dx.doi.org/10.1016/j.energy.2009.06.018]

[21] N. Abdoulmoumine, S. Adhikari, A. Kulkarni, and S. Chattanathan, "A review on biomass gasification syngas cleanup", *Appl. Energy,* vol. 155, pp. 294-307, 2015. [http://dx.doi.org/10.1016/j.apenergy.2015.05.095]

[22] A. Molino, S. Chianese, and D. Musmarra, "Biomass gasification technology: The state of the art overview", *J. Energy Chem.,* vol. 25, no. 1, pp. 10-25, 2016. [http://dx.doi.org/10.1016/j.jechem.2015.11.005]

[23] D.F. Call, *Development of a scalable microbial electrolysis cell and investigations of exoelectrogenic pure and mixed communities,* 2011.

[24] V. Claude, C. Courson, M. Köhler, and S.D. Lambert, "Overview and essentials of biomass gasification technologies and their catalytic cleaning methods", *Energy Fuels,* vol. 30, no. 11, pp. 8791-8814, 2016.
[http://dx.doi.org/10.1021/acs.energyfuels.6b01642]

[25] J.L. Wertz, and O. Bédué, *Lignocellulosic Biorefineries.* EPFL Press: Lausanne, Switzerland, 2013, pp. 1-521.
[http://dx.doi.org/10.1201/b15443]

[26] M. Rashidi, N. Alavi, F. Amereh, M. Rafiee, N. Amanidaz, K. Partovi, S. Mosanefi, and R. Bakhshoodeh, "Biohydrogen production from co-digestion of sugarcane vinasse and bagasse using anaerobic dark fermentation", *Bioresour. Technol. Rep.,* vol. 25, p. 101793, 2024.
[http://dx.doi.org/10.1016/j.biteb.2024.101793]

[27] Z. Fang, R.L. Smith, X. Qi, Ed., *Production of hydrogen from renewable resources.* vol. Vol. 5. Springer: Dordrecht, 2015.
[http://dx.doi.org/10.1007/978-94-017-7330-0]

[28] P.R. Lanjekar, N.L. Panwar, C. Agrawal, T. Gupta, K.K. Meena, G.L. Meena, and S.S. Meena, *A review on environmental assessment of thermochemical conversion technologies for energy generation from crop residue.* Earth Energy Sci, 2024.
[http://dx.doi.org/10.1016/j.ees.2024.07.001]

[29] I.B. Task, *Progress in Commercialization of Biojet/Sustainable Aviation Fuels (SAF).* Technologies and Policies, 2024.

[30] Z. Lian, Y. Wang, X. Zhang, A. Yusuf, L. Famiyeh, D. Murindatalaga, and Y. Sun, *Hydrogen Production by Fluidized Bed Reactors: A Quantitative Perspective Using the Supervised Machine Learning Approach*, 2021.

[31] C. Saffron, *Achieving Carbon-Negative Bioenergy Systems from Plant Material.*, 2020, p. xix–389.

[32] A. Blasi, A. Verardi, C.G. Lopresto, S. Siciliano, and P. Sangiorgio, "Lignocellulosic agricultural waste valorization to obtain valuable products: an overview", *Recycling,* vol. 8, no. 4, p. 61, 2023.
[http://dx.doi.org/10.3390/recycling8040061]

[33] M. Faizan, and H. Song, "Critical review on catalytic biomass gasification: State-of-Art progress, technical challenges, and perspectives in future development", *J. Clean. Prod.,* vol. 408, p. 137224, 2023.
[http://dx.doi.org/10.1016/j.jclepro.2023.137224]

[34] M.K. Singla, J. Gupta, R. Kumar, M. Safaraliev, M.H. Alsharif, A.A. Aly, and M.K. Kim, "A multi-model evaluation of Enhanced Tunicate Swarm Optimization for parameter identification", *Energy Rep.,* vol. 12, pp. 2049-2061, 2024.
[http://dx.doi.org/10.1016/j.egyr.2024.08.015]

[35] R. Narayan, M. Graziani, and P. Fornasiero, *Renewable Resources and Renewable Energy,* 2006.

[36] E. Kırtay, "Recent advances in production of hydrogen from biomass", *Energy Convers. Manage.,* vol. 52, no. 4, pp. 1778-1789, 2011.
[http://dx.doi.org/10.1016/j.enconman.2010.11.010]

[37] H.M.U. Ayub, M. Nizami, M.A. Qyyum, N. Iqbal, A.H. Al-Muhtaseb, and M. Hasan, "Sustainable hydrogen production *via* microalgae: Technological advancements, economic indicators, environmental aspects, challenges, and policy implications", *Environ. Res.,* vol. 244, p. 117815, 2024.
[http://dx.doi.org/10.1016/j.envres.2023.117815] [PMID: 38048865]

[38] M.R.G. Pangestu, Z. Malaibari, A. Muhammad, F.N. Al-Rowaili, and U. Zahid, "Comprehensive review on methane pyrolysis for sustainable hydrogen production", *Energy Fuels,* vol. 38, no. 15, pp.

13514-13538, 2024.
[http://dx.doi.org/10.1021/acs.energyfuels.4c01551]

[39] H.D. Setiabudi, M.A.A. Aziz, S. Abdullah, L.P. Teh, and R. Jusoh, "Hydrogen production from catalytic steam reforming of biomass pyrolysis oil or bio-oil derivatives: A review", *Int. J. Hydrogen Energy,* vol. 45, no. 36, pp. 18376-18397, 2020.
[http://dx.doi.org/10.1016/j.ijhydene.2019.10.141]

[40] R. Jain, N.L. Panwar, S.K. Jain, T. Gupta, C. Agarwal, and S.S. Meena, "Bio-hydrogen production through dark fermentation: an overview", *Biomass Convers. Biorefin.,* vol. 14, no. 12, pp. 12699-12724, 2024.
[http://dx.doi.org/10.1007/s13399-022-03282-7]

[41] N. Akhlaghi, and G. Najafpour-Darzi, "A comprehensive review on biological hydrogen production", *Int. J. Hydrogen Energy,* vol. 45, no. 43, pp. 22492-22512, 2020.
[http://dx.doi.org/10.1016/j.ijhydene.2020.06.182]

[42] G.D. Najafpour, M.H. Shahavi, and S.A. Neshat, "Assessment of biological Hydrogen production processes: A review", *IOP Conf. Ser. Earth Environ. Sci.,* vol. 36, no. 1, p. 012068, 2016.
[http://dx.doi.org/10.1088/1755-1315/36/1/012068]

[43] D. Das, and T.N. Veziroğlu, "Hydrogen production by biological processes: a survey of literature", *Int. J. Hydrogen Energy,* vol. 26, no. 1, pp. 13-28, 2001.
[http://dx.doi.org/10.1016/S0360-3199(00)00058-6]

[44] X.M. Guo, E. Trably, E. Latrille, H. Carrère, and J.P. Steyer, "Hydrogen production from agricultural waste by dark fermentation: A review", *Int. J. Hydrogen Energy,* vol. 35, no. 19, pp. 10660-10673, 2010.
[http://dx.doi.org/10.1016/j.ijhydene.2010.03.008]

[45] S. Dahiya, S. Chatterjee, O. Sarkar, and S.V. Mohan, "Renewable hydrogen production by dark-fermentation: Current status, challenges and perspectives", *Bioresour. Technol.,* vol. 321, p. 124354, 2021.
[http://dx.doi.org/10.1016/j.biortech.2020.124354] [PMID: 33277136]

[46] Y. Wang, Q.B. Zhao, Y. Mu, H.Q. Yu, H. Harada, and Y.Y. Li, "Biohydrogen production with mixed anaerobic cultures in the presence of high-concentration acetate", *Int. J. Hydrogen Energy,* vol. 33, no. 4, pp. 1164-1171, 2008.
[http://dx.doi.org/10.1016/j.ijhydene.2007.12.018]

[47] A.J. Percy, and M. Edwin, *A comprehensive review on the production and enhancement techniques of gaseous biofuels and their applications in IC engines with special reference to the associated performance and emission characteristics.* vol. Vol. 173087. Sci. Total Environ, 2024.

[48] G.P. Iragavarapu, S.S. Imam, O. Sarkar, S.V. Mohan, Y.C. Chang, M.V. Reddy, S-H. Kim, and N.K. Amradi, "Bioprocessing of waste for renewable chemicals and fuels to promote bioeconomy", *Energies,* vol. 16, no. 9, p. 3873, 2023.
[http://dx.doi.org/10.3390/en16093873]

[49] M.Y. Azwar, M.A. Hussain, and A.K. Abdul-Wahab, "Development of biohydrogen production by photobiological, fermentation and electrochemical processes: A review", *Renew. Sustain. Energy Rev.,* vol. 31, pp. 158-173, 2014.
[http://dx.doi.org/10.1016/j.rser.2013.11.022]

[50] C.N.C. Hitam, and A.A. Jalil, "A review on biohydrogen production through photo-fermentation of lignocellulosic biomass", *Biomass Convers. Biorefin.,* vol. 13, no. 10, pp. 8465-8483, 2023.
[http://dx.doi.org/10.1007/s13399-020-01140-y]

[51] G. Melitos, X. Voulkopoulos, and A. Zabaniotou, "Waste to sustainable biohydrogen production *via* photo-fermentation and biophotolysis: A systematic review", *Renewable Energy and Environmental Sustainability,* vol. 6, p. 45, 2021.
[http://dx.doi.org/10.1051/rees/2021047]

[52] J.B. McKinlay, and C.S. Harwood, "Photobiological production of hydrogen gas as a biofuel", *Curr. Opin. Biotechnol.,* vol. 21, no. 3, pp. 244-251, 2010. [http://dx.doi.org/10.1016/j.copbio.2010.02.012] [PMID: 20303737]

[53] S. Gupta, A. Fernandes, A. Lopes, L. Grasa, and J. Salafranca, "Photo-fermentative bacteria used for hydrogen production", *Appl. Sci. (Basel),* vol. 14, no. 3, p. 1191, 2024. [http://dx.doi.org/10.3390/app14031191]

[54] D.N. Dari, I.S. Freitas, F.I.D.S. Aires, R.L.F. Melo, K.M. dos Santos, P. da Silva Sousa, and J.C.D. Santos, "An updated review of recent applications and perspectives of hydrogen production from biomass by fermentation: A comprehensive analysis", *Biomass (Basel),* vol. 4, no. 1, pp. 132-163, 2024. [http://dx.doi.org/10.3390/biomass4010007]

[55] M. Li, P. Ning, Y. Sun, J. Luo, and J. Yang, "Characteristics and application of Rhodopseudomonas palustris as a microbial cell factory", *Front. Bioeng. Biotechnol.,* vol. 10, p. 897003, 2022. [http://dx.doi.org/10.3389/fbioe.2022.897003] [PMID: 35646843]

[56] G. Kumar, P. Bakonyi, G. Zhen, P. Sivagurunathan, L. Koók, S.H. Kim, G. Tóth, N. Nemestóthy, and K. Bélafi-Bakó, "Microbial electrochemical systems for sustainable biohydrogen production: Surveying the experiences from a start-up viewpoint", *Renew. Sustain. Energy Rev.,* vol. 70, pp. 589-597, 2017. [http://dx.doi.org/10.1016/j.rser.2016.11.107]

[57] V. Strezov, T.J. Evans, Ed., *Biomass Processing Technologies.* CRC Press, 2014. [http://dx.doi.org/10.1201/b17093]

[58] Andrea Papola, PhD Thesis, Università degli Studi di Napoli Federico II, 2017.,

[59] B. Senthil Rathi, P. Senthil Kumar, G. Rangasamy, and S. Rajendran, "A critical review on Biohydrogen generation from biomass", *Int. J. Hydrogen Energy,* vol. 52, pp. 115-138, 2024. [http://dx.doi.org/10.1016/j.ijhydene.2022.10.182]

[60] S. Lu, G. Tan, and X. Zhu, $H_2$ evolution catalysts for microbial electrolysis cells.*Novel Catalyst Materials for Bioelectrochemical Systems.* Fundamentals and Applications, 2020, pp. 27-43. [Editor(s)]

[61] M. Rosenbaum, M.A. Cotta, and L.T. Angenent, "Aerated Shewanella oneidensis in continuously fed bioelectrochemical systems for power and hydrogen production", *Biotechnol. Bioeng.,* vol. 105, no. 5, pp. 880-888, 2010. [http://dx.doi.org/10.1002/bit.22621] [PMID: 19998276]

[62] B.E. Logan, D. Call, S. Cheng, H.V.M. Hamelers, T.H.J.A. Sleutels, A.W. Jeremiasse, and R.A. Rozendal, "Microbial electrolysis cells for high yield hydrogen gas production from organic matter", *Environ. Sci. Technol.,* vol. 42, no. 23, pp. 8630-8640, 2008. [http://dx.doi.org/10.1021/es801553z] [PMID: 19192774]

[63] M. Sankir, N.D. Sankir, Ed., *Hydrogen Production Technologies.* John Wiley & Sons, 2017. [http://dx.doi.org/10.1002/9781119283676]

[64] A. Amrut Pawar, A. Karthic, S. Lee, S. Pandit, and S.P. Jung, "Microbial electrolysis cells for electromethanogenesis: Materials, configurations and operations", *Environ. Eng. Res.,* vol. 27, no. 1, p. 200484, 2022. [http://dx.doi.org/10.4491/eer.2020.484]

[65] K. Moustakas, M. Loizidou, M. Rehan, and A.S. Nizami, "A review of recent developments in renewable and sustainable energy systems: Key challenges and future perspective", *Renew. Sustain. Energy Rev.,* vol. 119, p. 109418, 2020. [http://dx.doi.org/10.1016/j.rser.2019.109418]

[66] L. Lu, D. Xing, and N. Ren, "Pyrosequencing reveals highly diverse microbial communities in microbial electrolysis cells involved in enhanced H2 production from waste activated sludge", *Water*

*Res.,* vol. 46, no. 7, pp. 2425-2434, 2012.
[http://dx.doi.org/10.1016/j.watres.2012.02.005] [PMID: 22374298]

[67] V. C. Kalia, "Microbes: Factories for Bioproducts", In: *Microbial Factories,* V. Kalia, Ed., Springer, 2015.

[68] M.B. Bahari, C.R. Mamat, A.A. Jalil, N.S. Hassan, M.H. Sawal, S. Rajendran, and M.N.H.Z. Alam, "Molybdenum as cathode materials: Paving the way for sustainable biohydrogen production in microbial electrolysis cells", *Process Saf. Environ. Prot.,* vol. 191, pp. 1633-1647, 2024.
[http://dx.doi.org/10.1016/j.psep.2024.09.032]

[69] R.N. Krishnaraj, R.K. Sani, Ed., *Bioelectrochemical Interface Engineering.* John Wiley & Sons, 2019.
[http://dx.doi.org/10.1002/9781119611103]

[70] M.K. Sarmah, T.P. Singh, P. Kalita, and A. Dewan, "Sustainable hydrogen generation and storage – a review", *RSC Advances,* vol. 13, no. 36, pp. 25253-25275, 2023.
[http://dx.doi.org/10.1039/D3RA04148D] [PMID: 37622026]

[71] R. Kumar, R. Singh, and S. Dutta, "Review and outlook of hydrogen production through catalytic processes", *Energy Fuels,* vol. 38, no. 4, pp. 2601-2629, 2024.
[http://dx.doi.org/10.1021/acs.energyfuels.3c04026]

[72] H. Thunman, M. Seemann, T. Berdugo Vilches, J. Maric, D. Pallares, H. Ström, G. Berndes, P. Knutsson, A. Larsson, C. Breitholtz, and O. Santos, "Advanced biofuel production via gasification – lessons learned from 200 man-years of research activity with Chalmers' research gasifier and the GoBiGas demonstration plant", *Energy Sci. Eng.,* vol. 6, no. 1, pp. 6-34, 2018.
[http://dx.doi.org/10.1002/ese3.188]

[73] P.C. Hallenbeck, and D. Ghosh, "Advances in fermentative biohydrogen production: the way forward?", *Trends Biotechnol.,* vol. 27, no. 5, pp. 287-297, 2009.
[http://dx.doi.org/10.1016/j.tibtech.2009.02.004] [PMID: 19329204]

[74] S. Kumar Gupta, S. Kumari, K. Reddy, and F. Bux, "Trends in biohydrogen production: major challenges and state-of-the-art developments", *Environ. Technol.,* vol. 34, no. 13-14, pp. 1653-1670, 2013.
[http://dx.doi.org/10.1080/09593330.2013.822022] [PMID: 24350426]

[75] M.I. Taipabu, K. Viswanathan, W. Wu, N. Hattu, and A.E. Atabani, "A critical review of the hydrogen production from biomass-based feedstocks: Challenge, solution, and future prospect", *Process Saf. Environ. Prot.,* vol. 164, pp. 384-407, 2022.
[http://dx.doi.org/10.1016/j.psep.2022.06.006]

[76] M. Balat, and M. Balat, "Political, economic and environmental impacts of biomass-based hydrogen", *Int. J. Hydrogen Energy,* vol. 34, no. 9, pp. 3589-3603, 2009.
[http://dx.doi.org/10.1016/j.ijhydene.2009.02.067]

[77] D.B. Levin, and R. Chahine, "Challenges for renewable hydrogen production from biomass", *Int. J. Hydrogen Energy,* vol. 35, no. 10, pp. 4962-4969, 2010.
[http://dx.doi.org/10.1016/j.ijhydene.2009.08.067]

[78] B. van Ruijven, D.P. van Vuuren, and B. de Vries, "The potential role of hydrogen in energy systems with and without climate policy", *Int. J. Hydrogen Energy,* vol. 32, no. 12, pp. 1655-1672, 2007.
[http://dx.doi.org/10.1016/j.ijhydene.2006.08.036]

[79] A. Demirbas, "Future hydrogen economy and policy", *Energy Sources B Econ. Plan. Policy,* vol. 12, no. 2, pp. 172-181, 2017.
[http://dx.doi.org/10.1080/15567249.2014.950394]

[80] B.D. Solomon, and A. Banerjee, "A global survey of hydrogen energy research, development and policy", *Energy Policy,* vol. 34, no. 7, pp. 781-792, 2006.
[http://dx.doi.org/10.1016/j.enpol.2004.08.007]

[81] K. Urbaniec, A. Friedl, D. Huisingh, and P. Claassen, "Hydrogen for a sustainable global economy", *J.*

*Clean. Prod.,* vol. 18, pp. S1-S3, 2010.
[http://dx.doi.org/10.1016/j.jclepro.2010.05.010]

[82] R.N. Krishnaraj, and J.S. Yu, *Systems biology approaches for microbial fuel cell applications.* vol. Vol. 125. Opportunities and Challenges, 2015.
[http://dx.doi.org/10.1201/b18718-12]

[83] P.K. Pandey, Z. El Khatib, R. Kumar, M. Singla, M. Safaraliev, and F. Kamalov, *A novel hybrid control strategy for grid-connected PV-battery system under different operating conditions*, 2024.
[http://dx.doi.org/10.15199/48.2024.09.56]

[84] M. Kumar Yadav, R. Kumar, R. Kumar Ratnesh, J. Singh, R. Chandra, A. Kumar, V. Vishnoi, G. Singh, and A. Kumar Singh, "Revolutionizing technology with spintronics: devices and their transformative applications", *Mater. Sci. Eng. B,* vol. 303, p. 117293, 2024.
[http://dx.doi.org/10.1016/j.mseb.2024.117293]

[85] P.K. Pandey, R. Kumar, V. Gupta, A. Kanungo, and S. Diwania, "An innovative hybrid controller-based combined grid-connected hybrid renewable energy system", *Electr. Eng.,* vol. 106, no. 6, pp. 7055-7072, 2024.
[http://dx.doi.org/10.1007/s00202-024-02363-2]

[86] Green Hydrogen: A Guide to Policy Making, International Renewable Energy Agency (IRENA), 2020. [Available from: https://www.irena.org/publications/2020/Nov/Green-hydrogen-A-guid--to-policy-making

[87] Hydrogen Program Plan, U.S. Department of Energy, 2020. [Available from: https://www.hydrogen.energy.gov/docs/hydrogenprogramlibraries/pdfs/hydrogen-program-plan-2020.pdf

[88] S. Madeddu, F. Ueckerdt, M. Pehl, J. Peterseim, M. Lord, K.A. Kumar, C. Krüger, and G. Luderer, "The $CO_2$ reduction potential for the European industry *via* direct electrification of heat supply (power-to-heat)", *Environ. Res. Lett.,* vol. 15, no. 12, p. 124004, 2020.
[http://dx.doi.org/10.1088/1748-9326/abbd02]

[89] *Hydrogen Insights 2021: A Perspective on Hydrogen Investment, Deployment and Cost Competitiveness,* 2021.https://hydrogencouncil.com/en/hydrogen-insights-2021/ Hydrogen Council

# CHAPTER 4

# Exploring Biomass: Opportunities and Innovations in Renewable Energy

**Nidhi Deepak Sharma[1,*], Kashmiri Khamkar[2], Nidhi Vijay Sharma[3]** and **Dipali K. Kolhe[1]**

[1] *Department of Chemistry, Jaywantrao Sawant College of Engineering Hadapsar, Pune 411028, Maharashtra, India*

[2] *Department of Chemistry, MIT World Peace University Kothrud, Pune 411038, Maharashtra, India*

[3] *Department of Chemistry, AISSMS's Institute of Information Technology, Pune 411001, Maharashtra, India*

**Abstract:** Biomass energy is an important approach for sustainable future development. This abstract explores the various biomass methods used for energy generation, emphasizing their environment values, technological improvements, and economic potential. A biomass energy system converts the organic waste, forestry byproducts, and agricultural wastes into usable energy. This process helps reduce environmental pollution and assists in waste management. Biomass energy sources include diverse types of wood waste, and byproducts from wood processing. In addition to that, there are some other biomass sources, including sugarcane molasses and residues from paper pulp processing. By incorporating these practices into our daily life and policy frameworks, societies can significantly reduce energy consumption and promote a cleaner and sustainable future. The widespread adoption of biomass energy is crucial for achieving long-term environmental and economic benefits, which ensure a robust and sustainable energy source. It sets alternative energy sources with conventional ones to focus on fossil fuel-related challenges, such as greenhouse gas emissions and global warming. This research examines various aspects of biomass energy, including its sources, conversion process, environmental impacts, and economic viability. This paper highlights the renewable energy technologies used to produce electricity and yield heat from naturally replenishing resources. The advancement in technology and policy support has made renewable energy more competitive with fossil fuels, upgrading energy security and promoting rural development in terms of cost and reliability.

**Keywords:** Biomass energy, Energy sources, Environment impact, Renewable energy.

* **Corresponding author Nidhi Deepak Sharma:** Department of Chemistry, Jaywantrao Sawant College of Engineering Hadapsar, Pune 411028, Maharashtra, India E-mail: drnidhi0927@gmail.com

**Reetu Malhotra, Pooja Mahajan, Manish Singla & Ramesh Kumar (Eds.)**

## INTRODUCTION

Biomass energy is an important component of all kinds of renewable and sustainable energy sources, which are obtained from the burning of organic materials derived from living things. Lignocellulosic biomass, which is obtained from plants or plant-based materials and is not used for food or feed, proposes a flexible and promising alternative [1]. Its wide range of sources and advanced conversion technologies contribute to reducing dependence on fossil fuels and addressing climate change. The biomass energy flow chart is depicted in Fig. (1).

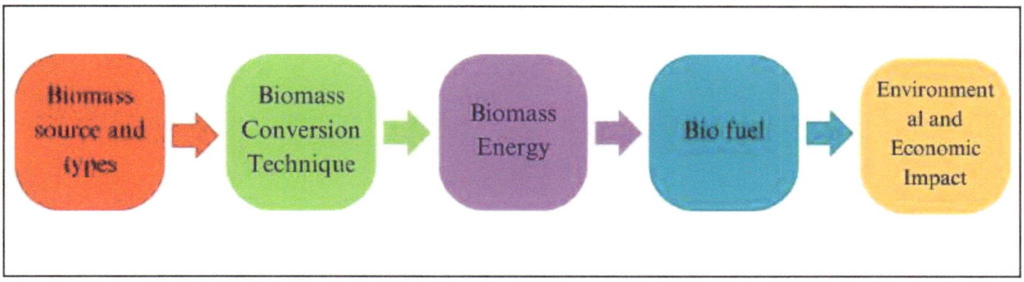


**Fig. (1).** Biomass Energy Flow Chart.

Bioenergy refers to energy harnessed from biomass materials and is currently the most extensively utilized renewable energy source globally [2]. Biomass resources are abundant and diverse, encompassing livestock manure, industrial and municipal waste streams (such as paper, plastics, and food waste), poultry byproducts, and sewage sludge. Bioenergy serves as an effective method for waste management, mitigating environmental pollution while contributing to economic stability [3]. This chapter explores the various forms of biomass, innovative technologies driving its development, and the opportunities and challenges associated with its use.

## TYPES OF BIOMASS AND THEIR SOURCES

Biomass can be categorized into different types based on its origin, composition, and characteristics [4]. Below are some key types of biomass, along with their respective sources, as shown in Fig. (**2**):

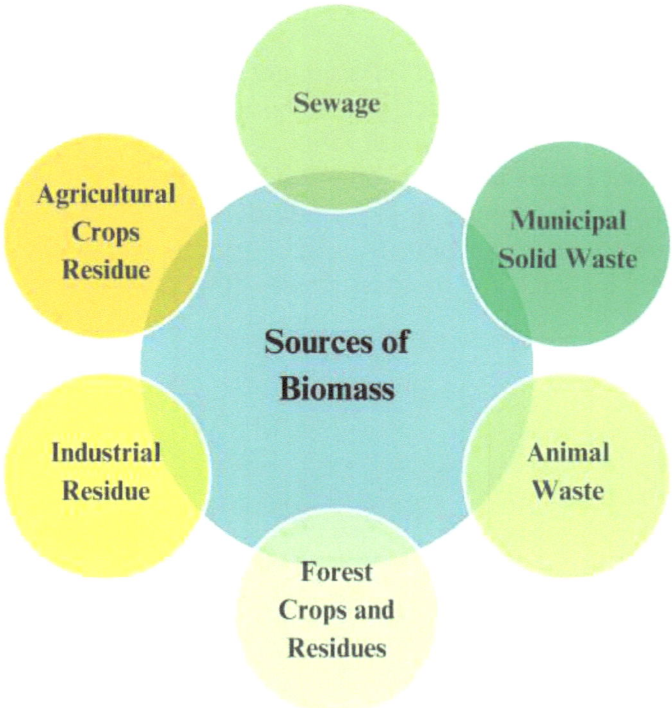


**Fig. (2).** Types of Biomass and their Sources [4].

- **Wood Biomass**
  Wood biomass originates from sources such as forest residues, wood chips, sawdust, and wood pellets. As one of the most traditional and extensively utilized forms of biomass, it can be directly combusted or processed into fuels. Using wood biomass can help reduce fossil fuel consumption, minimize greenhouse gas emissions, prevent deforestation, and support environmentally and economically sustainable practices [5].

- **Agricultural Residues**

Agricultural residues, including rice and wheat straw, bark, branches, corn husks, and fruit waste, are by-products of crop cultivation. These materials can be utilized directly as feedstock or converted into biomass products. Converting agricultural waste into biomass not only helps manage waste but also provides a range of sustainable alternatives for energy and material needs. The use of agricultural residues effectively enhances crop yields, improves soil fertility, provides additional income for farmers, and contributes to sustainability by recycling nutrients and organic matter [6].

- **Animal Manure**

Animal manure is produced by farm animals and is a valuable resource in agriculture due to its high nutrient content. By anaerobic digestion, manure can be converted into biogas. This process produces methane for energy and reduces the environmental impact of manure by minimizing methane emissions and producing a digestate that can be used as fertilizer, which is beneficial for plant growth [7].

- **Energy Crops**

Energy crops like switch grass, millet, wheat, starch crops, cellulosic crops, and sweet clover are cultivated for the purpose of producing energy, either through direct combustion or by being processed into biofuels. These crops are cultivated to provide renewable energy sources, thereby, helping to reduce the use of fossil fuels and contributing to environmental sustainability. However, their farming must be managed carefully to balance environmental, economic, and social factors, which can be utilized for electricity and heat generation, while the digestive can be used as a fertilizer [8].

- **Waste Biomass**

Biomass like paper mill sludge, municipal solid waste, food processing residues, energy crop residues, agricultural residues, forestry residues, and industrial organic by-products constitutes the waste biomass. These materials can be converted into energy through various technologies, helping to manage waste while generating power and heat. Proper management and utilization of waste biomass can help reduce waste, lower greenhouse gas emissions, and promote sustainability. Balancing economic and environmental considerations is key to maximizing the benefits of waste biomass [9].

- **Industrial Biomass**

Industrial waste is the organic waste from various industries such as paper mills, textile factories, metals, paints, sandpaper, slag, ash, and radioactive wastes. Through the refraction process, biomass is heated without oxygen at temperatures between 250°C and 350°C. The process removes moisture and volatile matter compounds, converting them into a solid product known as biochar or bio-coal, often used as a fuel for heat and energy generation or combined with coal in co-firing [10].

# BIOMASS CONVERSION TECHNOLOGIES

The primary pathways for bioenergy conversion technologies are presented in Fig. (**3**) [11].

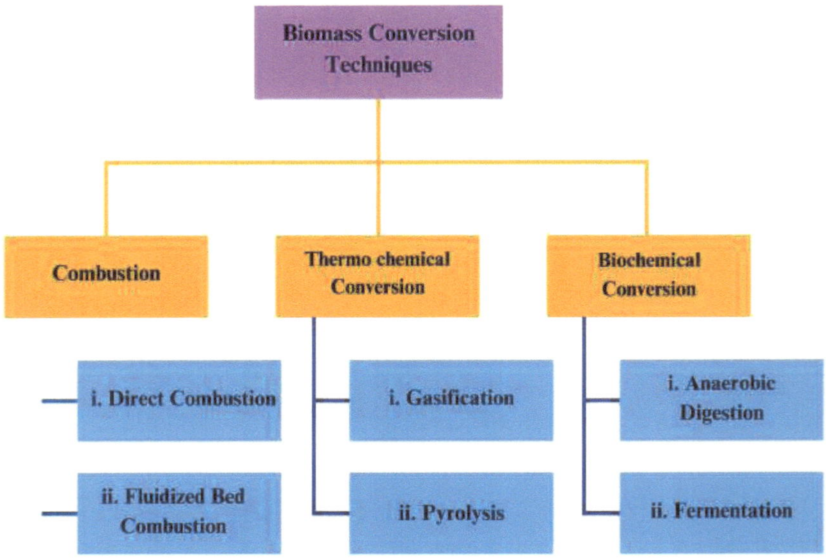


**Fig. (3).** Biomass Conversion Technologies [11].

## Combustion

Combustion is the most straightforward and oldest chemical process for converting biomass into energy. Biomass is burned in the presence of oxygen to generate heat, which can be harnessed for electricity production or district heating purposes. Modern combustion systems are equipped with technologies to reduce emissions and improve efficiency, by heating water to produce steam. After that, a turbine is driven by the steam. The turbine turns a shaft that links the turbine to a generator. The rotation of the shaft activates the generator, leading to the production of electricity [12].

### *Direct Combustion*

Direct combustion is the most straightforward method, where biomass is burned directly in a furnace or boiler to produce heat. This process is typically used for small to medium-sized applications [13].

a. Feedstock Preparation: Biomass is prepared by shredding, drying, and grinding.
b. Feeding: Biomass is fed into a combustion chamber.
c. Combustion Chamber: In this chamber, biomass is burned straight to generate heat.
d. Heat Transfer: Heat generated from the combustion is used to produce steam.
e. Emission Control: Filters and scrubbers are used to control discharges.
f. Ash Handling: At the end, the collected Ash is disposed of or recycled.

This technology is comparatively easy and well-established, and a variety of biomass feedstocks can be used. At present, numerous current biomass power plants and heating systems use direct combustion. However, combustion can produce air pollutants such as particulate matter, carbon monoxide, and nitrogen oxides, requiring effective emission control technologies. Processing and preparation of feedstock (such as drying and grinding) are also necessary for biomass to burn it efficiently [13].

### *Fluidized Bed Combustion*

Fluidized bed combustion involves suspending the biomass in a bed of hot, inert particles (commonly sand) that are fluidized by a stream of air. This method provides better control over the combustion process and enhances energy efficiency [14].

a. Feedstock Preparation: Biomass is prepared *via* a method similar to the direct combustion.
b. Feeding: Biomass is supplied into a fluidized bed combustor.
c. Fluidised Bed Combustor: Then biomass is burned in a bed of heated, fluidised particles.
d. Heat Transfer: The heat is extracted from the fluidised bed to generate steam..
e. Emission Control: Emission control is better because of more effective combustion.
f. Ash Handling: Ash is collected, which is almost in the same amount as in direct combustion, but normally easier to handle.

Fluidized bed combustion improves fuel flexibility and increases combustion efficiency, and also aids in lowering the quantity of environmental contaminants. Fuel flexibility promotes ecological sustainability, economic efficiency, and energy security.

However, more complex system designs, and higher capital costs are major challenges for this system. This can be used in the industrial boiler heating to reduce the utilization of fossil fuels. ompared with fluidised bed systems, direct combustion frequently has poorer efficiency as there is limited control over the combustion process and less optimisation of the burning conditions, which leads to reduced heat transfer efficiency. Fluidised bed combustion is better for bigger systems and gives more effective heat transmission, emission control, and monitoring; direct combustion is normally appropriate for smaller-scale applications. Both methods are applicable, and their uses depend on the efficiency, flexibility, heat transfer, and size of the operation [14].

## Thermochemical Conversion Technology

The thermochemical conversion methods include: i) Gasification, and ii) Pyrolysis.

### *Gasification*

The thermochemical process of gasification converts organic molecules into $CO_2$, $H_2$ & CO. This is achieved by adding just a little amount of steam or oxygen and using heat to react with organic molecules without complete combustion. The outcome is a mixture of gases, which are syngas, producer gas, or a mix of the two. Syngas can be used as a raw material for liquid fuels, to generate chemicals, or to produce electricity. As compared to direct combustion, gasification advances more usefulness and efficiency [15].

Based on the direction of gas flow through the fixed bed, they are divided into four categories: Updraft gasifier, Downdraft gasifier, Cross draft, and Fluidized bed Gasifier.

## APPLICATION:

a. Syngas can be used to produce electricity, chemicals, and synthetic fuels.
b. Syngas can be utilized in gas turbines or combined cycle power plants for electricity generation.
c. Syngas can be transformed into liquid fuels, such as synthetic diesel or gasoline, through processes like Fischer-Tropsch synthesis (Synthetic Fuels).
d. Syngas can be used to produce chemicals like methanol, ammonia, and hydrogen.
e. The process produces heat that can be utilized for industrial applications or district heating.
f. Converts agricultural residues, forestry residues, and other organic materials

into syngas for energy production and biofuels (Biomass Gasification).

g. It is a process that produces municipal solid waste or industrial waste to reduce volume and produce useful products (Waste Gasification).
h. It is used on a large scale in power plants to convert coal into syngas for cleaner energy production (Coal Gasification).

Gasification offers flexibility in processing a diverse range of feedstock, such as coal, biomass, and waste, and can be adjusted to various operational conditions. This process transforms feedstock into valuable products like syngas or electricity, boosting the technology's economic, environmental, and operational advantages, thus providing a more efficient and sustainable solution for energy production and waste management. Gasification processes can reduce certain emissions related to conventional combustion, especially when integrated with advanced cleaning and conversion technologies. Gasification efficiency rates vary widely based on technology, feedstock, and operational conditions, with cold gas efficiency typically ranging between 70-80%, fuel conversion rates reaching up to 100%, and IGCC systems demonstrating efficiencies from around 39.9% to 43.0%—though carbon capture reduces these figures. These numbers highlight the importance of optimizing gasification processes to maximize energy output while minimizing environmental impact, making gasification a crucial component in advancing sustainable energy solutions [16].

However, Gasification processes are complex and require careful management of reaction conditions and syngas clean-up. Gasification technology's initial capital investment is high.

Gasification can reduce emissions, but it still produces $CO_2$ and other pollutants, so observing effective management is challenging nowadays [15].

### *Pyrolysis*

Pyrolysis is a thermochemical process that decomposes organic material at high temperatures between 310°C to 900°C in the absence of oxygen. Because during this process, in the absence of oxygen, the organic matter doesn't burn, instead, it breaks down into three different forms: syngas, a solid known as biochar, and a liquid called bio-oil [15].

a. Biochar (or Charcoal): The solid residue left after pyrolysis, which can be used as a soil amendment, in filtration, or as a carbon-rich material for various applications.
b. Bio-oil (or Pyrolysis Oil): A complex liquid composed of various organic

compounds, which can serve as a fuel or be further refined into chemicals and other products.

c. Syngas: A gas mixture mainly consisting of carbon monoxide (CO), hydrogen ($H_2$), and carbon dioxide ($CO_2$), which can be utilized for energy generation or as a feedstock in chemical production.

Pyrolysis is used for various applications, including waste management, biofuel production, and materials processing.

a. Slow Pyrolysis: Performed at lower temperatures and slower heating rates, resulting in a higher yield of solid char. It is commonly used for producing biochar.
b. Fast Pyrolysis: Performed at higher temperatures and faster heating rates, focusing on maximizing the yield of bio-oil. It is used for producing liquid biofuels.
c. Flash Pyrolysis: An extremely fast pyrolysis process with very rapid heating rates, to produce bio-oil with less char formation.

In the pyrolysis process, waste materials such as plastics, tires, and other organic waste are converted into useful products and reduce landfill volume. Crop leaves, husk, bark, and other organic waste can be processed to produce biochar and other by-products. Bio-oil is used as a renewable fuel for producing chemicals and as an alternative biofuel. It serves as a source of electricity generation or as a raw material for producing synthetic fuels and chemicals. Activated carbon, like biochar produced through pyrolysis, can be utilized in filtration and purification processes. The pyrolysis process can produce carbon nanomaterials with applications in electronics and materials science [17].

Pyrolysis helps in reducing volume and impacts on the environment since it allows the useful elements of trash and biomass to be extracted. Pyrolysis systems can pyrolyze various feedstocks and be used under different operating conditions. Pyrolysis presents crucial benefits through lowering emissions, as it is a thermal degradative process and carried out in the absence of oxygen. However, the process may be complicated and expensive due to the high accuracy of temperature and reaction conditions required. The feedstock and process variables affect the quality and composition of products that can potentially influence their utility and value. Pyrolysis is a relatively energy-intensive process; hence, input and output control is necessary to sustain the operation. Treatment processes help to neutralize or safely eliminate any toxic or toxic substances, and leftovers and by-products should be handled in a manner that ensures environmental protection [18].

## Biochemical Process

### *Bio-Digestion*

Anaerobic digestion is a biological process that decomposes organic matter without oxygen, resulting in the production of biogas (mainly methane) and digestate, due to its efficient feedstock conversion and high selectivity. This process is commonly employed for organic waste management and renewable energy generation. The key steps in liquid biofuel production include the pre-treatment of lignocellulose feedstock (such as cellulose and lignin), biological or enzymatic hydrolysis, fermentation of the resulting sugars by microorganisms, and the extraction of bioethanol, which is an environmentally friendly fuel [19].

i. Feedstock Preparation: Organic Material: Inputs can include agricultural residues, food waste, manure, sewage sludge, and other organic materials. The feedstock may be pre-processed (e.g., shredded or mixed) to enhance the digestion process [20].
ii. Digestion Stages: (a) Hydrolysis: Complex organic compounds, such as carbohydrates, proteins, and fats, are broken down into simpler molecules (sugars, amino acids, and fatty acids) by hydrolytic bacteria. (b) Acidogenesis: Simple molecules are transformed into volatile fatty acids, hydrogen, and carbon dioxide by acidogenic bacteria. (c) Acetogenesis: Volatile fatty acids and other by-products are further converted into acetic acid, along with additional hydrogen and carbon dioxide, by acetogenic bacteria. (d) Methanogenesis: Methanogenic bacteria convert acetic acid and hydrogen into methane ($CH_4$) and carbon dioxide ($CO_2$), generating biogas [20].
iii. Biogas Collection - Composition: Biogas primarily consists of methane (50-70%) and carbon dioxide (30-50%), with trace amounts of other gases like hydrogen sulphide ($H_2S$) and ammonia ($NH_3$) [21].
iv. Digested Management: Solid and Liquid Fractions: The residual material left after the digestion process is known as digested matter. It is composed of a solid fraction that is rich in fibrous and organic matter, along with a liquid fraction that is nutrient-rich.

Anaerobic digestion is used to process municipal solid waste, food waste, and agricultural residues, reducing landfill use and managing organic waste efficiently. This is also used in wastewater treatment plants to process sewage sludge, reducing its volume and producing renewable energy. The methane produced during anaerobic digestion can be used for generating electricity, heating, or upgraded to biomethane for use as a renewable vehicle fuel. The nutrient-rich digestate serves as a valuable fertilizer or soil conditioner, enhancing soil fertility and reducing the need for synthetic fertilizers [19].

Biogas can be used for electricity generation, heating, or as a vehicle fuel after purification and upgrading. Moreover, biogas is a renewable source of energy that can replace fossil fuels and help to greenhouse gas emissions. It helps control and reduce the amount of organic waste that is usually landfilled. The recycling process converts organic waste into useful digestate, acting like a nutrient-rich soil amendment and thereby lessening the need for chemical fertilisers. In addition to producing useful by-products like digestate and biogas that can be used, anaerobic digestion can reduce foul odours produced during the breakdown of organic waste [21].

However, the feedstock's composition may also affect the quality of the biogas produced and the efficiency of the digestion process. Temperature, pH, retention time, and other variables must be monitored and controlled during the anaerobic digestion process. The economic feasibility of anaerobic digestion depends on several factors, which include feedstock prices, expenditures for infrastructure, and the market value of the biogas. Safe exploitation of biogas depends on proper management and treatment to avoid pollution release [20].

### *Fermentation*

The fermentation process for the formation of biogas primarily involves the breakdown of organic materials by microorganisms like bacteria, yeast, or fungi in an anaerobic (oxygen-free) environment, which is converted into sugars and other carbohydrates, such as alcohol, acids, and gases. Ethanol can serve as a future fuel source for powering automobiles. For instance, bioethanol is produced from the fermentation of sugars derived from crops such as corn or sugarcane [22]. Pharmaceuticals and chemical antibiotics such as penicillin are also produced by fermentation processes involving fungi or bacteria, and vitamins and amino acids are used as supplements and food additives. Waste Treatment during biogas production causes fermentation of organic waste in anaerobic digesters, which produces biogas (methane) and syngas for energy [23]. In this way, fermentation uses renewable resources to sustainably produce biofuels and chemicals.

However, during the process, sterile conditions are maintained to control unwanted microorganisms. To obtain the best results from fermentation, we need to carefully control factors such as temperature and pH to help the process run smoothly. Fermentation creates leftover materials and waste, so it is important to handle or recycle these by-products properly. The cost of ingredients, managing the process, and scaling up production can affect whether fermentation is a cost-effective method for making products [23].

## INNOVATIONS IN BIOMASS (ADVANCED BIOFUELS)

Advanced biofuels are next-generation fuels derived from renewable biological resources that go beyond traditional biofuels like ethanol and biodiesel [24]. These technologies promise to reduce competition in food production and enhance sustainability.

- **Source:** Burning biomass is one of the ways to release its energy. The main resources of biomass energy are: (i) Animal manure and human sewage: They are converted into biogas, primarily methane, which can be used as a fuel. (ii) Agricultural crops and waste materials: They are either burned as fuel or fermented to produce liquid biofuels like ethanol. (iii) Food and wood waste in garbage: They are burned for electricity generation in power plants or converted into biogas (methane) in landfills. (iv) Forest residues and wood processing waste: They are burned to provide heating for buildings, produce industrial process heat, and generate electricity [25].
- **Production Processes**: This process comprises drying, pyrolysis, combustion, char gasification, and cracking [25]. (i) Hydrothermal Liquefaction (HTL): In this process, biomass is exposed to high pressure and temperature in the presence of water, resulting in the production of bio-crude oil [26] (iia) Gasification Process: In this process, biomass is converted into syngas. Syngas is a mixture of hydrogen, carbon monoxide, and carbon dioxide, which can be used to produce biofuels through chemical synthesis [27]. (iii) Anaerobic Digestion Process: Organic materials are decomposed by microorganisms in the absence of oxygen to produce biogas (methane and carbon dioxide) [19]. (iv) Transesterification: Process: Oils and fats are converted into biodiesel and glycerol by reacting with alcohol (usually methanol or ethanol) in the presence of a catalyst [28].
- **Advantages of Biomass Energy**: Biomass has several merits why it can be regarded an excellent substitute for fossil fuel [29]:
    i. Among other forms of renewable energy, it is unique as it offers a promising future.
    ii. Ecosystem activity is not negatively affected.
    iii. Biomass energy is global in nature and is unrestricted.
    iv. The cost of biomass energy is lower than other energy.
    v. Biomass energy lowers dependency on grids.
    vi. Biomass energy reduces the waste buried at landfills.
- **Disadvantages of Biomass Energy:** Though biomass energy has its benefits, it does have some drawbacks as well. The disadvantages of biomass energy include [29]:

i. Biomass as an energy source does not remain environmentally clean.
ii. There will be a threat of many forests being cut down.
iii. Quite a lot of freshwater resources would be essential.
i. Biomass is not as effective as fossil fuels.

A more versatile and self-sufficient system may be developed by combining biomass with other renewable resources, such as solar or wind. There are also potential developments in new energy technologies, including waste-to-energy systems, which would help treat urban and industrial waste into usable fuel.

## DIFFERENTIATING BIOMASS AND OTHER TYPES OF RENEWABLE ENERGY SOURCES

This section describes the differences between biomass and other resources as energy sources [30].

a. Biomass *vs.* Solar and Wind Energy

**Biomass:**

i. **Source:** Organic materials like plants and waste.
ii. **Reliability:** Produces energy continuously, regardless of weather.
iii. **Land Use:** Uses land for growing crops or managing waste.
iv. **Emissions:** Releases $CO_2$ but less than fossil fuels; cause some pollution.

**Solar and Wind Energy:**

i. **Source:** Sunlight (solar) and wind.
ii. **Reliability:** Energy output varies with weather and time of day.
iii. **Land Use:** Requires space, but can be placed on rooftops or in unused areas.
iv. **Emissions:** Very low during operation, but some impact from manufacturing.

b. Biomass *vs.* Fossil Fuels

**Biomass:**

i. **Source:** Organic materials (plants, waste).
ii. **Emissions:** Lower $CO_2$ than fossil fuels; less pollution.
iii. **Renewable:** It is renewable as long as resources are managed well.
iv. **Impact:** Can use land and produce some waste.

**Fossil Fuels:**

i. **Source:** Coal, oil, natural gas.
ii. **Emissions:** High $CO_2$ and pollutants; major environmental impact.
iii. **Renewable:** Non-renewable; they deplete over time.
iv. **Impact:** Significant environmental damage and health risks.

## PROS AND CONS FOR BIOMASS ENERGY

### (a). Pros:

i. Uses Waste: Can turn organic waste into energy.
ii. Less Pollutant: Produces less sulphur and mercury compared to coal.
iii. Cost-Effective: Often cheaper, especially if using waste materials.
iv. Reduces Landfill Waste: Helps cut down on waste that goes to landfills.
v. Local Production: Can be made locally, reducing the need for large power plants.
vi. Versatile: Can be used to make fuels, plastics, and medicines.

### (b). Cons:

i. $CO_2$ Emissions: Burning still emits carbon dioxide into the atmosphere.
ii. Pollution: Can potentially generate foul smells and pollutants.
iii. Deforestation Risk: It may cause deforestation if used on a large scale.
iv. Irrigation Needs: Irrigation requirements will be increased.
v. Less Efficient: The efficiency of biomass energy is less compared to other energy sources.

## CHALLENGES AND BARRIERS

a. Technical Challenges

Biomass conversion technologies are still at an early development stage despite their potential for commercialization; they want to focus on some technical issues such as efficiency, scale-up up and design, which are feedstock dependent [31].

b. Economic Barriers

Widespread use of biomass technologies is hindered because biomass energy is more expensive than fossil fuels, has higher investment costs, with restricted sustainability of such technologies. There is a need to transform the idea of independent policies into appropriate government financing support and investor expectations [32].

c. Environmental Concerns

While biomass energy has several environmental advantages, several problems remain, including land use change, potential negative effects on biodiversity, and emissions arising from some of the conversion processes [31].

## CONCLUSION

Biomass energy is undoubtedly one of the most sustainable resources that can be used for a diverse range of applications. As explored in this chapter, advancements in technology continue to enhance the efficiency and viability of biomass as a key player in the global energy transition. From agricultural residues to plant-based biofuels, the diversification of biomass sources opens new opportunities for energy security and environmental stewardship. However, challenges such as land use concerns, carbon emissions management, and economic feasibility must be carefully navigated to ensure biomass fulfills its potential as a sustainable solution. The present technology overcomess the challenges of using biomass as a source of renewable and sustainable energy. With ongoing research and policy support, the future of biomass energy promises to integrate cutting-edge innovations, creating a balanced approach to meeting energy demands while preserving ecological integrity. Embracing these opportunities can pave the way for a cleaner, more resilient energy future. Additionally, ongoing research, providing assistance, and collaborating on a global scale are crucial strategies for maximizing the potential of biomass energy and making a positive impact on the environment, economy, and society.

## REFERENCES

[1] J.A. Okolie, S. Nanda, A.K. Dalai, and J.A. Kozinski, "Chemistry and specialty industrial applications of lignocellulosic biomass", *Waste Biomass Valoriz.,* vol. 12, no. 5, pp. 2145-2169, 2021. [http://dx.doi.org/10.1007/s12649-020-01123-0]

[2] W.V. Reid, M.K. Ali, and C.B. Field, "The future of bioenergy", *Glob. Change Biol.,* vol. 26, no. 1, pp. 274-286, 2020. [http://dx.doi.org/10.1111/gcb.14883] [PMID: 31642554]

[3] M.K. Awasthi, S. Sarsaiya, A. Patel, A. Juneja, R.P. Singh, B. Yan, S.K. Awasthi, A. Jain, T. Liu, Y. Duan, A. Pandey, Z. Zhang, and M.J. Taherzadeh, "Refining biomass residues for sustainable energy and bio-products: An assessment of technology, its importance, and strategic applications in circular bio-economy", *Renew. Sustain. Energy Rev.,* vol. 127, p. 109876, 2020. [http://dx.doi.org/10.1016/j.rser.2020.109876]

[4] R. Malhotra, and B. Singh, "Design and construction of solar shed", *AIP Conference Proceedings,* vol. 2220, p. 140061, 2020.

[http://dx.doi.org/10.1063/5.0001157]

[5] P. Lauri, P. Havlík, G. Kindermann, N. Forsell, H. Böttcher, and M. Obersteiner, "Woody biomass energy potential in 2050", *Energy Policy,* vol. 66, pp. 19-31, 2014. [http://dx.doi.org/10.1016/j.enpol.2013.11.033]

[6] H. Unal, and K. Alibas, "Agricultural residues as biomass energy", *Energy Sources B Econ. Plan. Policy,* vol. 2, no. 2, pp. 123-140, 2007. [http://dx.doi.org/10.1080/15567240600629401]

[7] C. Sasikumar, R. Sundaresan, M. Nagaraja, and C. Rajaganapathy, "A review on energy generation from manure biomass", *Mater. Today Proc.,* vol. 45, pp. 2408-2412, 2021. [http://dx.doi.org/10.1016/j.matpr.2020.10.832]

[8] R.H. Sims, A. Hastings, B. Schlamadinger, G. Taylor, and P. Smith, "Energy crops: current status and future prospects", *Glob. Change Biol.,* vol. 12, no. 11, pp. 2054-2076, 2006. [http://dx.doi.org/10.1111/j.1365-2486.2006.01163.x]

[9] E. Iakovou, A. Karagiannidis, D. Vlachos, A. Toka, and A. Malamakis, "Waste biomass-to-energy supply chain management: A critical synthesis", *Waste Manag.,* vol. 30, no. 10, pp. 1860-1870, 2010. [http://dx.doi.org/10.1016/j.wasman.2010.02.030] [PMID: 20231084]

[10] Z. Anwar, M. Gulfraz, and M. Irshad, "Agro-industrial lignocellulosic biomass a key to unlock the future bio-energy: A brief review", *Journal of Radiation Research and Applied Sciences,* vol. 7, no. 2, pp. 163-173, 2014. [http://dx.doi.org/10.1016/j.jrras.2014.02.003]

[11] P. Adams, T. Bridgwater, A. Lea-Langton, A. Ross, and I. Watson, Biomass conversion technologies.*Greenhouse Gas Balances of Bioenergy Systems.* Academic Press, 2018, pp. 107-139. [http://dx.doi.org/10.1016/B978-0-08-101036-5.00008-2]

[12] B.M. Jenkins, L.L. Baxter, and J. Koppejan, *"Biomass combustion," Thermochemical Processing of Biomass: Conversion into Fuels.* Chemicals and Power, 2019, pp. 49-83. [http://dx.doi.org/10.1002/9781119417637.ch3]

[13] M. Mandø, Direct combustion of biomass.*Biomass Combustion Science, Technology and Engineering.* Woodhead Publishing, 2013, pp. 61-83. [http://dx.doi.org/10.1533/9780857097439.2.61]

[14] R. Chirone, P. Salatino, F. Scala, R. Solimene, and M. Urciuolo, "Fluidized bed combustion of pelletized biomass and waste-derived fuels", *Combust. Flame,* vol. 155, no. 1-2, pp. 21-36, 2008. [http://dx.doi.org/10.1016/j.combustflame.2008.05.013]

[15] S. Jha, J.A. Okolie, S. Nanda, and A.K. Dalai, "A review of biomass resources and thermochemical conversion technologies", *Chem. Eng. Technol.,* vol. 45, no. 5, pp. 791-799, 2022. [http://dx.doi.org/10.1002/ceat.202100503]

[16] V. Kirubakaran, V. Sivaramakrishnan, R. Nalini, T. Sekar, M. Premalatha, and P. Subramanian, "A review on gasification of biomass", *Renew. Sustain. Energy Rev.,* vol. 13, no. 1, pp. 179-186, 2009. [http://dx.doi.org/10.1016/j.rser.2007.07.001]

[17] Z. Yu, H. Ma, X. Liu, M. Wang, and J. Wang, "Review in life cycle assessment of biomass conversion through pyrolysis-issues and recommendations", *Green Chem. Eng.,* vol. 3, no. 4, pp. 304-312, 2022. [http://dx.doi.org/10.1016/j.gce.2022.08.002]

[18] G. Wang, Y. Dai, H. Yang, Q. Xiong, K. Wang, J. Zhou, Y. Li, and S. Wang, "A review of recent advances in biomass pyrolysis", *Energy Fuels,* vol. 34, no. 12, pp. 15557-15578, 2020. [http://dx.doi.org/10.1021/acs.energyfuels.0c03107]

[19] D. Fabbri, and C. Torri, "Linking pyrolysis and anaerobic digestion (Py-AD) for the conversion of lignocellulosic biomass", *Curr. Opin. Biotechnol.,* vol. 38, pp. 167-173, 2016. [http://dx.doi.org/10.1016/j.copbio.2016.02.004] [PMID: 26948108]

[20] S. Paul, and A. Dutta, "Challenges and opportunities of lignocellulosic biomass for anaerobic digestion", *Resour. Conserv. Recycling,* vol. 130, pp. 164-174, 2018. [http://dx.doi.org/10.1016/j.resconrec.2017.12.005]

[21] B. Bharathiraja, T. Sudharsana, J. Jayamuthunagai, R. Praveenkumar, S. Chozhavendhan, and J. Iyyappan, "RETRACTED: Biogas production–A review on composition, fuel properties, feed stock and principles of anaerobic digestion", *Renew. Sustain. Energy Rev.,* vol. 90, pp. 570-582, 2018. [http://dx.doi.org/10.1016/j.rser.2018.03.093]

[22] Y. Lin, and S. Tanaka, "Ethanol fermentation from biomass resources: current state and prospects", *Appl. Microbiol. Biotechnol.,* vol. 69, no. 6, pp. 627-642, 2006. [http://dx.doi.org/10.1007/s00253-005-0229-x] [PMID: 16331454]

[23] P.C. Munasinghe, and S.K. Khanal, "Biomass-derived syngas fermentation into biofuels: Opportunities and challenges", *Bioresour. Technol.,* vol. 101, no. 13, pp. 5013-5022, 2010. [http://dx.doi.org/10.1016/j.biortech.2009.12.098] [PMID: 20096574]

[24] S. Arbogast, D. Bellman, J.D. Paynter, and J. Wykowski, "Advanced bio-fuels from pyrolysis oil: The impact of economies of scale and use of existing logistic and processing capabilities", *Fuel Process. Technol.,* vol. 104, pp. 121-127, 2012. [http://dx.doi.org/10.1016/j.fuproc.2012.04.036]

[25] M. González Prieto, M. Fortunatti Montoya, P.E. Hegel, and S. Pereda, "Supercritical reactors for the production of advanced bio-fuels: A review", *J. Supercrit. Fluids,* vol. 134, pp. 106-113, 2018. [http://dx.doi.org/10.1016/j.supflu.2017.11.020]

[26] B. Biswas, J. Kumar, and T. Bhaskar, Advanced hydrothermal liquefaction of biomass for bio-oil production.*Biofuels: Alternative Feedstocks and Conversion Processes for the Production of Liquid and Gaseous Biofuels.* Academic Press, 2019, pp. 245-266. [http://dx.doi.org/10.1016/B978-0-12-816856-1.00010-5]

[27] R. Thomson, P. Kwong, E. Ahmad, and K.D.P. Nigam, "Clean syngas from small commercial biomass gasifiers; a review of gasifier development, recent advances and performance evaluation", *Int. J. Hydrogen Energy,* vol. 45, no. 41, pp. 21087-21111, 2020. [http://dx.doi.org/10.1016/j.ijhydene.2020.05.160]

[28] R. Estevez, L. Aguado-Deblas, F.J. López-Tenllado, C. Luna, J. Calero, A.A. Romero, F.M. Bautista, and D. Luna, "Biodiesel is dead: Long life to advanced biofuels—A comprehensive critical review", *Energies,* vol. 15, no. 9, p. 3173, 2022. [http://dx.doi.org/10.3390/en15093173]

[29] S.V. Vassilev, C.G. Vassileva, and V.S. Vassilev, "Advantages and disadvantages of composition and properties of biomass in comparison with coal: An overview", *Fuel,* vol. 158, pp. 330-350, 2015. [http://dx.doi.org/10.1016/j.fuel.2015.05.050]

[30] G.V. Ochoa, J.N. Alvarez, and C. Acevedo, "Research evolution on renewable energies resources from 2007 to 2017: A comparative study on solar, geothermal, wind and biomass energy", *International Journal of Energy Economics and Policy,* vol. 9, no. 6, pp. 242-253, 2019. [http://dx.doi.org/10.32479/ijeep.8051]

[31] N. Banerjee, "Biomass to energy—an analysis of current technologies, prospects, and challenges", *BioEnergy Res.,* vol. 16, no. 2, pp. 683-716, 2023. [http://dx.doi.org/10.1007/s12155-022-10500-7]

[32] D. Mignogna, M. Szabó, P. Ceci, and P. Avino, "Biomass energy and biofuels: perspective, potentials, and challenges in the energy transition", *Sustainability (Basel),* vol. 16, no. 16, p. 7036, 2024. [http://dx.doi.org/10.3390/su16167036]

# CHAPTER 5

# Renewable Energy Forecasting using Advanced Machine Learning Techniques

**Himanshi**[1], **Gaganpreet Kaur**[1,*], **Jatin Arora**[1], **Senthil Kumar AV**[2], **Ujjwal Kaushik**[3] and **Chander Prabha**[1]

[1] *Department of Computer Science & Engineering, Chitkara University Institute of Engineering and Technology, Chitkara University, Rajpura 140401, Punjab, India*

[2] *Department of Computer Applications, Hindusthan College of Arts and Sciences, Udayampalayam 641028, Tamil Nadu, India*

[3] *Chitkara College of Pharmacy, Chitkara University, Rajpura 140401, Punjab, India*

**Abstract:** The integration of renewable energy sources into the power grid presents significant challenges due to their recurrent and variable nature. Accurate forecasting of renewable energy generation is crucial for grid stability, efficient energy management, and the optimisation of energy resources. This chapter explores the application of advanced machine learning techniques in renewable energy forecasting, which focuses on solar and wind energy. It begins with an introduction to the importance of renewable energy forecasting and the role of machine learning in addressing forecasting challenges. The fundamentals of renewable energy forecasting are discussed, highlighting the key challenges and the necessity for accurate predictions. The chapter delves into various machine learning techniques, including regression, classification, clustering, deep learning, ensemble methods, and hybrid models. A comprehensive comparison between traditional and advanced machine learning techniques is provided. Hybrid models and ensemble methods are explored for their potential to enhance forecasting accuracy. Real-time forecasting and optimisation techniques are also covered, emphasising their applications in smart grids and microgrids. The chapter includes real-world case studies and applications, showcasing successful implementations of machine learning in renewable energy forecasting. The ethical and social implications of using AI and ML in this domain are discussed, along with policy and regulatory considerations. This chapter aims to provide a comprehensive overview of the state-of-the-art in renewable energy forecasting using advanced machine learning techniques, offering valuable insights for researchers, practitioners, and policymakers.

**Keywords:** Autoregressive Integrated Moving Average (ARIMA), Cloud Motion Vectors (CMV), Data privacy, Emerging technologies, Numerical Weather Prediction (NWP), Renewable Energy Sources (RES), Security concern.

[*] **Corresponding author Gaganpreet Kaur:** Department of Computer Science & Engineering, Chitkara University Institute of Engineering and Technology, Chitkara University, Rajpura 140401, Punjab, India; E-mail: drgaganpreetcse@gmail.com

**Reetu Malhotra, Pooja Mahajan, Manish Singla & Ramesh Kumar (Eds.)**

# INTRODUCTION

As demand for renewable energy increases, it is imperative to invest in research and development of renewable energy sources [1, 2]. Renewable energy is essential for reducing greenhouse gas emissions and mitigating severe climate change [3 - 5]. There are many advantages of renewable energy sources, such as reduced dependence on foreign energy, increased employment, and potential cost savings [1, 6]. Renewable energies include those that come directly from the sun, such as thermal, photochemical, and photoelectric energy, or indirectly, like wave energy, wind power, and the biomass energy stored in organic matter [7]. Renewable energy sources are continuously replenished by nature on a daily basis [7].

## Types of Renewable Energy Sources

Energy plays a vital role in driving a nation's economic growth and overall development, serving as a fundamental input for industrial, agricultural, and technological progress. Renewable natural resources are the primary source of energy. Renewable energy sources are energy sources derived from the environment that are sustainable. Renewable energy is a never-ending source of clean energy and a vast source to replace the increasing need for fossil fuels. The major advantage of renewable energy over non-renewable energy is that they are readily available, can be renewed easily, and has a long lifespan. Nuclear power, biomass, geothermal, wind, solar, tidal, and wave energy are a few of them. If utilized efficiently, each of these energy sources has the potential to provide reliable and compatible energy while minimizing the burden on the environment. Renewable energy, derived from natural resources that replenish themselves over regular cycles, offers a sustainable and eco-friendly alternative to conventional fossil fuels. (Fig. **1**) illustrates the diagrammatic representation of various types of renewable energy sources.

- **Solar energy:** It is derived directly from sunlight and is the most abundant and frequently utilized energy source. Solar power is converted into thermal or electrical energy using mirrors and photovoltaic panels. It is the most widely available and cleanest renewable energy. Although sunlight availability varies by location and weather conditions, solar energy remains one of the most convenient and beneficial energy solutions for both domestic and industrial applications.
- **Wind energy:** Wind energy is energy derived from moving air. Large onshore and offshore turbines are used to harness the kinetic energy of moving air to generate electricity. Recent technological advancements, such as the development of larger rotor diameters and taller turbines, have significantly

improved power output and efficiency. Despite the intermittent nature of wind, many regions around the world possess substantial potential to develop wind energy as a reliable renewable source.

- **Hydro energy:** Hydroelectricity is generated using the energy of water currents. This is generated using the movement between differences in elevation, such as when water descends *via* reservoirs or rivers. Besides providing electric energy, hydropower reserves also provide navigation services and irrigation, as well as drinking and agricultural supplies. Currently, the most widespread form of renewable electricity is generated through hydropower. A common approach that is based on the regular patterns of rainfall could be affected by droughts when other techniques, including ecosystem changes, come into play.
- **Biomass:** Bioenergy is produced from biomass, which includes organic materials such as wood, charcoal, manure, and crops. It is common in rural areas of developing countries for heating, lighting, and cooking. Finally, some examples of current biomass sources include crops dedicated to bioenergy (second generation), agricultural and forestry residues, and organic waste. However, the exploitation of energy can lead to significant land-use changes, including desertification consequences due to deforestation. Therefore, their usage should be limited as much as necessary.
- **Geothermal energy:** Geothermal power is generated when heat is available from the Earth's interior through techniques like drilling or other methods. The surface geothermal fluids have the potential to generate power. Hydrothermal reservoir electricity generation is one of the oldest, most reliable technologies in energy production for more than 100 years.
- **Nuclear power:** Nuclear power is the production of electricity using heat produced from nuclear fission in a reactor. A nuclear power plant operates similarly to a large coal-fired power plant, with the key difference being the heat source: the nuclear reactor replaces the boiler. Aside from that, it uses the same essential components—pumps, valves, turbines, generators, steam generators, condensers, and other necessary instruments.

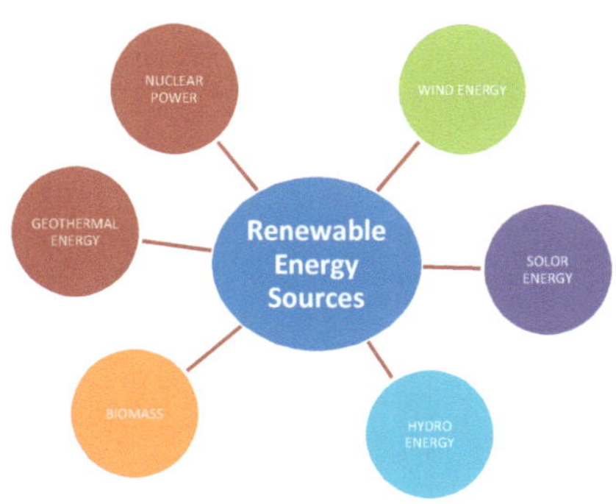


**Fig. (1).** Types of Renewable Energy Sources.

## Importance of Renewable Energy Forecasting

The projected amount of electricity generated from renewable sources has a great impact on power system administration and operation. The accuracy of renewable energy generation is crucial to ensure the grid is dependable and reliable in its delivery of electricity to consumers, thereby reducing the risks, adverse effects, and costs associated with the energy market and business platforms [8]. One of the most crucial factors affecting the weather is that it impacts all types of renewable energy sources, despite its volatile nature, which constantly makes it unpredictable to a certain degree. Renewable Energy Sources (RES), due to their inherent large-scale variations and disparities, are challenging to manage and integrate into the electrical system [9]. One way to address this particular problem is perhaps by developing sound projection models through the manufacturing of renewable energy. Efficient forecasting models may help mitigate the negative effects of the unpredictability and uncertainty associated with energy from renewable resources being discharged to the electrical network. Energy forecasting has historically been conducted using traditional modelling and analysis techniques, including mathematical and physical calculations [10]. However, statistical models that use the Autoregressive Integrated Moving Average (ARIMA) approach have limitations because renewable energy data has a high-dimensional structure and also due to complex nonlinear relationships [11].

Various models, such as UV radiation-based physical models and Numerical Weather Prediction (NWP) models, are commonly used to estimate renewable energy. Parameters commonly used for weather forecasting, such as solar radiation, come in cloud cover and other atmospheric features, while features for forecasting wind power and direction are considered as atmospheric parameters. Nevertheless, these models are not up to the mark due to the numerous factors associated with the Earth's atmosphere and the inherent complexities of weather prediction. Additional analysis and enhancement are needed to improve the precision of these models and overcome these problems [12, 13].

## Machine Learning Techniques in Energy Forecasting: An Overview

Machine learning aims to equip the robot with data abstraction capabilities and enhance its execution of specific tasks. It is the construction of models that allows computers to classify data or utilize those classifications to predict or decide action. With the help of ML, the computer is trained to recognise and counter particular types of inputs by providing it with an array of instances or 'training data'. Through this, the computer gains the potential to identify patterns and make decisions or predictions based on unseen information that has been taught to it [14 - 16]. The applications of machine learning are numerous, particularly in areas such as finance, safety, and insurance, healthcare, and e-commerce. Likewise, systems for renewable energy can be more effectively controlled by determining how much renewable energy can be produced with the aid of ML algorithms. There are many machine learning algorithms present today, and each of them has its benefits and drawbacks as well. One needs to differentiate between the following primary types of algorithms. There are three types of artificial intelligence programs: supervised training, unsupervised machine learning, and reinforcement learning [17]. The classifications of machine learning and the techniques available are given in Fig. (**2**) below.

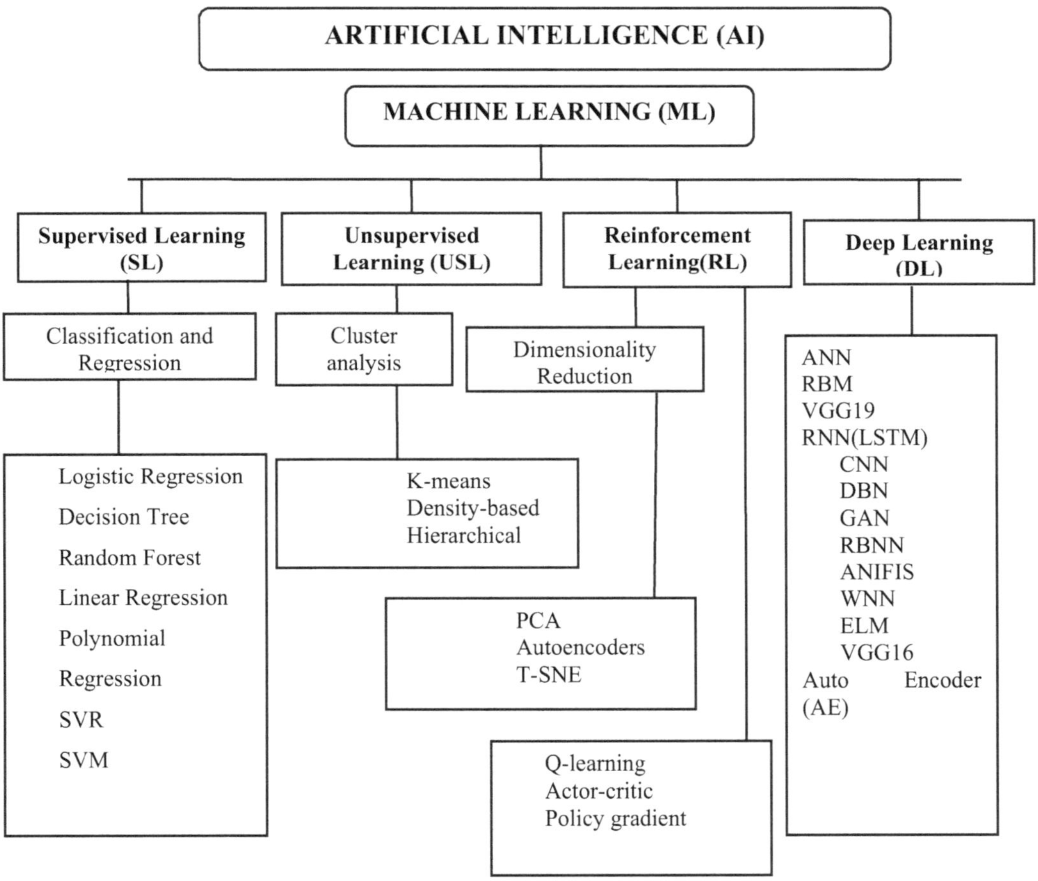


**Fig. (2).** Machine Learning types and available algorithms.

## FUNDAMENTALS OF RENEWABLE ENERGY FORECASTING

Renewable energy forecasting is the process of predicting the future amount of energy that will be generated by renewable energy sources, such as water, solar, and wind-emitting facilities. Such forecasts are typically issued for a short period.

Thus, the primary purpose of using environmentally friendly methods for energy forecasting is to provide timely and accurate information on the amount of electricity to be produced from different sources. To make the most of the current available energy resources, this data is needed by grid operators and energy administrators to schedule energy dispatch to meet supply and demand. The theory of accurate expectations is a prerequisite for the eradication of blackouts and provides continuous confirmation that there is always enough energy to meet consumption in the electricity grid.

## Key Challenges in Forecasting Renewable Energy

- **Unpredictable Events:** Unexpected disturbances, notably catastrophic weather conditions and fluctuations, particularly in the form of disruptions, remain erratic, and include natural calamities, disasters and human-instigated events in the energy market. It was difficult to forecast production with renewable energy accurately because of its high sensitivity to various factors, which ranged from price changes to availability and general market demands. The means of addressing these risks should be designed to ensure that the availability of services remains steady despite the observed trend of constantly rising initial costs.
- **Increased Complexity:** Integrating decentralized and renewable energy sources into power networks presents a certain level of complexity. Many of these sources exhibit random or cyclical behavior, leading to challenges such as grid stability and cybersecurity risks. As a result, they demand specialized management strategies and increased attention to prevent disruptions to organizational operations.
- **Digital Transformation:** There is a need to transform administration as more electric power systems become increasingly intricate. It is made possible through artificial intelligence, which also enhances security and compliance, and enables the installation of smart meters. The problems that current energy networks and distribution systems present to energy firms can be more effectively addressed with the use of technologies such as artificial intelligence.
- **Changing Customer Expectations**: Traditionally used utility models are currently experiencing significant changes with the emergence of consumers who generate their own power. Influenced by culture, demographic characteristics, and value processes that reflect higher levels of autonomy and individualization, power influences a shift in energy providers' offerings. To meet new and current requirements of consumers, provide unique services, and analyze usage rates in real-time, new systems need to be incorporated.
- **Decentralised Power Generation:** This results in the decentralisation of power in many ways, often as a nominal consequence of the shift to renewable energy. The decentralized government may lead to increased expenses and make

forecasting a bit more difficult. Utilities must design complex statistical models that can help predict demand in conjunction with supply within a less clear context, in order to manage localized power generation.

### Importance of Forecasting for Grid Stability and Energy Management

Climate prediction procedures play a crucial role in the optimal integration of renewable energy into electrical systems. To be more effective, they must be both accurate and timely. These days, grid operators can predict shifts in power supply and consumption by using information from meteorological radar, satellite imagery, and other advanced observation facilities.

Current Numerical Weather Prediction (NWP) models have significantly enhanced the predictability of renewable electricity output. These models provide overall projections of expected output from sources such as solar and wind energy by incorporating factors that include the aerodynamics of wind flow, cloud formation, and solar irradiation.

The system managers may improve the level of integration of renewable energy into the electricity grid and guarantee the efficiency of the power supply using the actual meteorological data and NWP outputs. Furthermore, the grid integrator is also able to process huge volumes of meteorological data at a faster pace due to the advancements in AI and holistic data analysis tools. These digital tools help to observe weather patterns and trends, and improve the grid stability and accuracy of clean electricity production.

## MACHINE LEARNING TECHNIQUES FOR FORECASTING

Machine Learning (ML) has emerged as a powerful tool for forecasting in renewable energy systems due to its ability to process and analyze large and complex datasets [18]. The goal of Machine Learning (ML) forecasting approaches is similar to that of traditional methods- to improve prediction accuracy while reducing the mistakes, which are mostly symbolised by the term 'loss'. Machine learning comes with the feature of analyzing large and complex data, as it aims to identify relationships and features that cannot be recognised by statistical methods. Machine learning models process real-time and historical data to dynamically adapt to the present conditions to capture more accurate and responsive predictions. As opposed to traditional methods that can sometimes necessitate a periodical recalibration of the algorithm, the ML algorithms constantly acquire new data and changes in patterns. This adaptive ability enables the models to revise the previous probabilities and enhance prediction reliability with time.

Moreover, the flexibility of ML methods, in turn, broadens their usage to diverse fields of forecasting, such as the output of renewable energy, trends in financial markets, and weather changes, optical conditions, which are often associated with volatility and uncertainty. Therefore, machine learning is not only more accurate and predictive, but it also creates a robust and resistant structure to deal with the issues that are typically linked to dynamic and data-intensive forecasting problems.

## Overview of Machine Learning Algorithms

Some of the machine learning algorithms are:

### *Regression*

Regression is a supervised machine learning algorithm that aims at inferring a continuous target variable from a set of predictor variables [18]. Regression aims to identify a mathematical model that maps inputs to a continuum of outputs, which can be singular or range [19]. Regression is another machine learning technique used to predict a continuous outcome variable that depends on one or more predictor variables. The primary goal of regression methods is to identify a specific mathematical equation that defines the relationship between the target variable and the features. After this function is defined, it can then be applied to data that have not been used to build the function to make the prediction. Some of the commonly used regression algorithms include linear regression, polynomial regression, and support vector regression algorithms, all of which have unique methods for determining the best model for the data. Linear regression aims to model data with a straight line, while polynomial regression fits the data using a curve. In contrast, support vector regression focuses on maximizing the margin between the data points and the regression line. Using an adequate regression analysis, one can identify intricate patterns and forecast within a desired range in numerous fields, such as financial or medical.

- **Polynomial and Linear Regression:** Based on one or more input variables, linear regression is actually one of the simplest forms of forecasting a continuous output variable [18]. This relationship of input and output variables is depicted by a straight line [20]. On the other hand, nth-degree polynomial functions are applied in polynomial regression, a type of linear regression to explain how the input factors are related to the dependent variable [21].
- **Support Vector Regression (SVR):** The main goal of Support Vector Machines (SVMs) is to define an ideal choice line, or margin, to achieve the greatest gap between data sets, and therefore enhance the accuracy of classification and the reduction of the gap between the estimated and actual values. SVMs can process both linear and non-linear data, and convert input

variables into a high-dimensional feature space with the help of kernel functions. This is what allows SVMs to be effective in specifically predicting the renewable energy output, where the data is usually intricate and affected by many variable environmental factors. SVMs can also be used to make renewable energy production predictions more accurate and believable by detecting small patterns and trends in meteorological and operational data [18].

### *Classification*

Classification is a supervised learning method that is well-suited for use in problems where the output is set in categorical classification. The main premise of classification is to define an input-output relationship in which the input variables are matched with these well-defined discrete classes. This makes it particularly useful in situations where it is necessary to estimate in which category a new observation should be classified. In the process of modeling future performance and output of Renewable Energy Sources (RES), logistic regression, decision trees, random forests, and Support Vector Machines (SVMs) are some of the most commonly used methods. These models are used to classify the various states of the system, such as high or low energy production, depending on the current conditions, such as weather parameters, past performance record and equipment conditions.

- Logistic Regression: This process calculates the probability that an output variable is true or false using a sigmoid function. In renewable energy forecasting, it is capable of making statements about the occurrence and non-occurrence of specific events, such as the power generation level of a solar or wind farm.
- Decision Trees: These are a different kind of categorization, which function based on subdividing the input space into finer parts depending on the values of the input variables and thereafter assigning each of these portions a label or a value [22].
- Random Forest: A common and accurate prediction model is the random forest, which falls under the supervised learning approach. This approach involves creating numerous decision trees and aggregating their results [23].
- Support Vector Machines (SVM): They are classification algorithms that seek a hyperplane with the maximum margin between the hyperplane and the data points, similar to Support Vector Regression (SVR) [24, 25].

### *Cluster Analysis*

Cluster analysis is an approach in the unsupervised learning process in which data are divided into suitable subgroups of similar records based on the characteristics of the data. Unlike supervised learning, which has the data labelled in advance to help the model refine its predictions, cluster analysis does not require labels to be set at any one time. Instead, it defines natural clusters within the data, which makes it possible to uncover new patterns within that data. The aim of clustering is to divide data into clusters so that data points within the same cluster are closer to each other. This approach can be applied in different areas, like market segmentation, facial recognition, and social network studies, where one would find it beneficial to discern the natural, unorganized way of data. Thus, by identifying these clusters, cluster analysis allows for the further identification of meaningful patterns and relations within the complex set of data, thereby providing better support for decision-making and a deeper understanding of the matter.

- K-means: The technique involves mapping every data point to the nearest cluster centroid, and the next centroid is defined as the average of the next cluster [26, 27].
- There is a set of algorithms named hierarchical clustering, where the cluster set is partitioned and merged iteratively according to the distance or similarity of its elements, and the result resembles a tree. The other class of clustering algorithm is the density-based, which operates by clustering data points that are densely populated and isolating them from regions of low density [26 - 28].

## COMPARISON BETWEEN TRADITIONAL VS. ADVANCED MACHINE LEARNING TECHNIQUES

Table **1** presents the comparison between traditional and available advanced machine learning techniques.

**Table 1. Traditional vs. Advanced Machine Learning Techniques.**

| Field of comparison | Traditional Machine Learning Techniques | Advanced Machine Learning Techniques |
|---|---|---|
| **Intricacy** | Use simpler models like simple decision trees and linear regression. | Use ensemble approaches, including random forests and more complex models like neural networks that are able to deal with more complex issues. |
| **Achievement** | Useful for such datasets as small datasets, structured datasets, and datasets that have a distinct pattern. | Learn better on large, unstructured and complex data like text, images and sensor data. |
| **Readability** | Readable and interpretable, so that the users can clearly interpret and understand the process of making the decision. | In many cases, often called black-box models because of their extreme complexity, it is hard to understand the decision-making process. |

| Field of comparison | Traditional Machine Learning Techniques | Advanced Machine Learning Techniques |
|---|---|---|
| **Scalability** | Cannot solve complex problems or even work with very large data. | They have been made comfortable in dealing with high and massive data flows, complex scenarios included. Designed to scale smoothly. |
| **Adaptability** | Based on manual feature engineering and transformations, which provide little adaptation to raw data. | Very adaptable and able to automatically learn patterns and features of raw data with a very small amount of handwork. |

## SOLAR ENERGY FORECASTING

The increasing consumer need, limitations, and expensive nature of traditional sources of energy like petroleum and coal and the inexhaustible supply of solar energy have provided a good chance for a solution to the world energy crisis. This change is provoking tremendous progress in the global energy industry [29]. The diversity of solar energy utilization in different sectors of the energy system leads to a significant improvement in the stability, sustainability, and effectiveness of the energy systems in the modern world [30].

Proper estimation of solar power production needs a thorough knowledge of solar and atmospheric conditions, radiation dispersion pattern and the operation principle of solar plants. This knowledge will be necessary not only to ensure the successful operation of power grids but also to improve energy trading approaches and guarantee the optimal operation of solar power plants [31].

### Techniques for Solar Irradiance Prediction

- **Statistical Tactics:**

Historical data analysis involves examining past meteorological and solar irradiance data to identify procedural patterns and detect trends relevant to renewable energy forecasting. By examining the long-term data, the researchers will be able to identify the connection between climatic variables and the behaviour of the solar radiation. The level of solar irradiance in the future can be forecasted using statistical techniques such as linear regression and time series forecasting models, including Autoregressive Integrated Moving Average (ARIMA) models, which utilize past data to predict future data.

Advantages: Less calculation power is needed; it is easy for users to understand.

Cons: May struggle to capture complex, nonlinear patterns or sudden variations in weather conditions; Accuracy can decline when applied to highly dynamic or unpredictable environmental scenarios.

- **Numerical Weather Prediction (NWP):**

Physical Models: NWP systems try to imitate the real-world process, for instance, solar radiation, using a numerical approximation. For the purpose of estimating solar irradiance, these models utilize other factors, such as temperature, humidity, cloud type, and wind conditions.

Advantages: It provides detailed, accurate predictions in line with the atmosphere's actual mechanics.

Cons: This requires high-quality input data and computationally demanding algorithms.

- **Methods Based on Satellites:**

Satellite imagery: Satellites can get pictures of the Earth's surface and the atmosphere, which can be used to determine the amount of solar radiation, the amount of cloud cover or any other atmospheric condition. With this data, it is also possible to approximate sun irradiation at different locations.

Advantages: It is useful for big predictions and immediately reflects events and covers all aspects.

Cons: Interpretation is a challenging process, and data resolution may or may not be high.

- **Hybrid Methods:**

Here, multiple prediction methods are employed. As such, this strategy involves leveraging the strengths of each method in order to improve the overall prediction capability of the system.

Pros: It enlists the benefits of several techniques by integrating them to enhance the performance of prediction.

Cons: Implementation may have to be complex, while tuning may require the development of complex strategies.

- **Analysis of Cloud Motion Vectors (CMV):**

Cloud mapping is a short-term method that predicts small fluctuations in sunlight intensity with the help of cloud movement and their behaviour, which is identified *via* satellite imagery or pictures taken from the ground level. Solar Irradiance changes with the difference in cloud cover. (Fig. **3**) illustrates the techniques for predicting solar irradiance.

Advantages: It provides exact, limited-scope predictions of variations of irradiance.

Cons: May not offer as much value in terms of long-term trends.

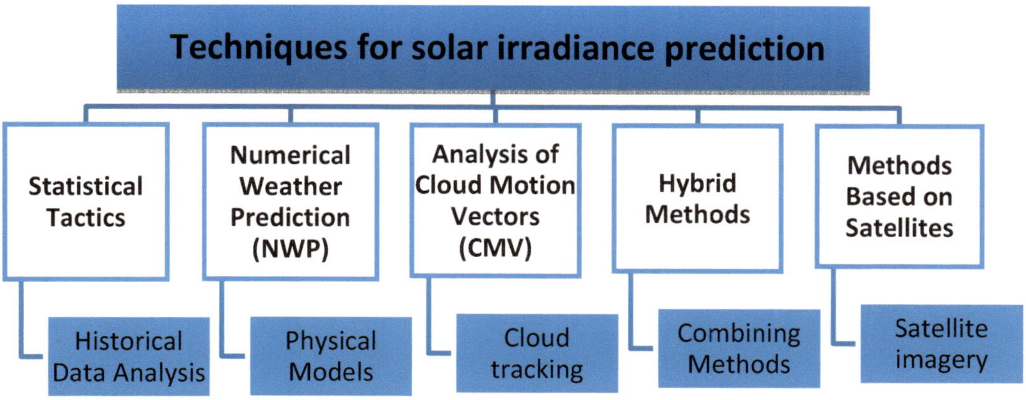


**Fig. (3).** Techniques for Solar Irradiance Prediction.

## Case Studies and Applications

- **The European Solar Radiation Atlas (ESRA)**

The ESRA clear sky model is established for calculating the solar radiation on the Earth's surface, employing satellite images with the Heliosat method [32]. The proposed ESRA model is far more accurate and robust than benchmark models on account of incorporating the Linke Turbidity Factor (T_L)and site elevation, which are crucial for accurate estimation of solar radiation [32].

Application: The ESRA clear-sky model is applied for the precise evaluation of the solar radiation from the satellite images, which is essential for the efficiency control of the solar systems and the forecast [32].

- **Solar Radiation at the National Renewable Energy Laboratory (NREL)**

Comparison with observed solar radiation at three sites in the north-eastern U. S. shows that NREL overestimates low radiation levels by using total sky cover [33].

Application: Higher details in estimating solar radiation by including the cloud base height coupled with visibility-obstructing factors have the potential to improve precision, especially in lesser radiance conditions applicable in energy prediction and climate simulation [33].

- **California Independent System Operator's Solar Forecasting (CAISO)**

However, solar forecasting has made great improvements, but many of the models are not feasible for Independent System Operators (ISOs) [34]. For the hourly data forecasts, it employs the North American Mesoscale (NAM) forecast system and then employs the rapid similarity search for enhancing the data into fifteen-minute intervals, suitable for the lead-time and resolution [34].

Application: This solar forecasting algorithm is critical to CAISO since it provides accurate and timely information needed for better management of the energy grid and integration of renewable energy systems [34].

- **Solar Forecasting and the Australian Renewable Energy Agency (ARENA)**

Detailed solar forecasting is provided by ARENA based on satellite images, numerical meteorological models for solar irradiance and power generation, incorporated with machine learning techniques for different solar forecasting projects across Australia. Due to the growing deployment of renewable sources of energy, such as wind and solar energy, a precise forecast of these renewable sources' output remains critical to the advancement of the Australian power industry [35].

Application: Improving methodologies for forecasting, such as filling in the flaws, such as conditionally changing variance and oversizing of solar fields, is amongst the most vital for grid operation and efficient incorporation of renewable energy into the system [35 - 37].

## PERFORMANCE METRICS AND EVALUATION

Various performance metrics and evaluation methods are discussed in this section.

### Performance Metrics:

- **Mean Absolute Error (MAE):** It determines the mean deviation from the actual values of the predicted values or *vice versa* of the set values. The evaluation style is relatively simple compared to the other two methods, as it simply determines to what extent forecasts deviate from actual values irrespective of the direction.
- **Root Mean Square Error (RMSE)**: RMSE is very much like MAE, but in this kind of error, large errors are given more weight since, after averaging, the values are squared [38, 39]. This is why RMSE is even more valuable, as it shows larger errors and gives information about the quality of the forecasts by focusing on and showing higher mistakes that potentially could be critical.

- **Mean Absolute Percentage Error (MAPE):** The MAPE translates the prediction error into a percentage of the actual values, making the results more interpretable and allowing for comparison with different levels of renewable energy generation. This metric is useful in a scenario where one wants to compare the forecast performance between datasets of different scales.
- **R-Squared ($R^2$):** The coefficient of determination, represented by the symbol $R^2$ estimates the amount of variance of the actual data that can be accounted for by the model. A higher $R^2$ value also means that a greater amount of deviation in the data is explained by the model. Hence, it has better predictive ability.
- **Normalized Mean Bias Error (NMBE):** It quantifies the level of bias of the model in the prediction by expressing the average error as a percentage of the mean of the observed data. It favours identifying systematic biases because it points out whether the model under- or overestimates the target variable systematically.
- **Skill Score:** The skill score evaluates an advanced forecasting model against a more basic model, such as an average or naive forecast model. They compare the accuracy that the new model provides against the old model and provide a measure of how much better the advanced model is than the simple baseline.

**Evaluation Methods:**

- **Cross-Validation:** K-fold cross-validation is a method that assesses the performance of a given model by dividing the data set into distinct samples called folds, where the process of training and testing takes place in sequential order. Cross-validation is a method in which the data segments are rotated between the model to make sure that the model works for a given portion of the dataset not in the entire section of the data segment. Subsequently, this helps verify the general applicability and reliability of the model.
- **Temporal Resolution Analysis:** Prognostic accuracy evaluation assesses a model's ability to predict results within a specific duration, such as every hour, day, or month, which is known as temporal resolution analysis. This is important in applications that may need to use different forecasting techniques for different horizons, such as renewable power generation forecasting. Regarding how the model performs across various time scales, it can be understood that when used for different forecast situations, the needs are met as required.
- **Scenario Testing:** Scenario testing requires checking the model on different settings, weather conditions, levels of the energy produced and other such circumstances that may arise. This way it makes the model flexible and capable of responding to different situations, as this is very much needed, especially in volatile fields like energy forecasting, where changes can occur in a very short time.

- **Comparative Analysis:** This kind of evaluation technique measures the difference between an enhanced model and the standard forecasting techniques to assess general performance enhancements. Comparing this advanced model with others also reveals these improvements, and therefore shows how well the advanced model performs by having such a benchmark.
- **Outlier Detection:** Outlier detection focuses on the prediction values, which vary significantly from the actual values. This way, it is possible to identify the causes of prediction inaccuracies in specific cases by investigating these outliers. This is a crucial step in tuning the model, optimizing forecast accuracy, and ensuring it is ready for any situation it may encounter in the future.

## FUTURE DIRECTIONS

- **Integration of Transfer Learning:** Use the experience gained from the well-populated regions for the data to increase performance in underrepresented areas for forecasts [40].
- **Standardised Benchmarking:** Develop and standardize the datasets, evaluation techniques, and standards for conducting transparent and reproducible benchmarks on the forecasting models [41].
- **Hybrid Modelling Approaches:** Combine physical theories with machine learning methodologies while maximizing forecasting by combining knowledge of underlying processes with practical data interpretation.
- **Real-time Adaptive Models:** Design systems to adeptly update predictions as more data streams through and to adapt to seasonal changes in weather as well as energy demand.
- **Multi-source Data Integration:** Aggregate data from satellites, Internet of Things devices and weather services to augment inputs and strengthen predictions of models.

## LIMITATIONS

In addition to the advancement of new machine learning methodologies that have significantly improved the performance of renewable energy anticipation predictions, much remains to be done. The accuracy of the model depends heavily on the nature and amount of historical data that may be limited or unreliable in some areas. Also, because many such machine learning models lack transparency, understanding the factors driving prediction in such models continues to present considerable challenges. Furthermore, existing models fail to adapt easily to different geographies/weather patterns and thus require massive retraining before deployment in unrelated environments. Furthermore, the use of complex models like deep learning requires enormous computations, and this can make real-time applications problematic in regions with scanty computational resources.

## CONCLUSION

In this chapter, the most emphasized resource is the renewable one, which outlines the possibilities for solving the existing problems associated with climate change and energy consumption. While the context of growing demand for clean energy solutions, the aforesaid price volatility, and stable forecasting of renewable energy generation is of particular importance. The conventional types of approaches in forecasting are somewhat helpful in a broad sense, but they fail to deal precisely with the intricacies of renewable energy sources such as solar and wind. This is where machine learning comes in handy with better tools that can withstand big, complex data and perform in the presence of condition changes.

Various aspects of machine learning are discussed with examples such as regression, classification, and clustering, to illustrate how they can be used to better predict renewable energy outputs. Thus, it can enhance grid balance and efficiency in energy control, and consequently facilitate the integration of renewable power into the system.

All in all, the chapter aims to demonstrate that the combination of renewable energy and machine learning is a fascinating field of research and practice. In this way, humans move towards the future where renewable energy is not an addition to fossil energy sources, but is a dominant part of the energy supply.

## REFERENCES

[1] D. Gielen, F. Boshell, D. Saygin, M.D. Bazilian, N. Wagner, and R. Gorini, "The role of renewable energy in the global energy transformation", *Energy Strategy Reviews,* vol. 24, pp. 38-50, 2019. [http://dx.doi.org/10.1016/j.esr.2019.01.006]

[2] W. Strielkowski, L. Civín, E. Tarkhanova, M. Tvaronavičienė, and Y. Petrenko, "Renewable energy in the sustainable development of electrical power sector: A review", *Energies,* vol. 14, no. 24, p. 8240, 2021. [http://dx.doi.org/10.3390/en14248240]

[3] G.A. Tiruye, A.T. Besha, Y.S. Mekonnen, N.E. Benti, G.A. Gebreslase, and R.A. Tufa, "Opportunities and challenges of renewable energy production in Ethiopia", *Sustainability (Basel),* vol. 13, no. 18, p. 10381, 2021. [http://dx.doi.org/10.3390/su131810381]

[4] N.E. Benti, T.A. Woldegiyorgis, C.A. Geffe, G.S. Gurmesa, M.D. Chaka, and Y.S. Mekonnen, "Overview of geothermal resources utilization in Ethiopia: Potentials, opportunities, and challenges", *Sci. Am.,* vol. 19, p. e01562, 2023.

[5] N.E. Benti, A.B. Aneseyee, C.A. Geffe, T.A. Woldegiyorgis, G.S. Gurmesa, and M. Bibiso, "Biodiesel production in Ethiopia: Current status and future prospects", *Sci. Am.,* vol. 19, p. e01531, 2023.

[6] N.E. Benti, Y.S. Mekonnen, and A.A. Asfaw, "Combining green energy technologies to electrify rural community of Wollega, Western Ethiopia", *Sci. Am.,* vol. 19, p. e01467, 2023.

[7] D.O. Hall, and J.I. Scrase, "Will biomass be the environmentally friendly fuel of the future?", *Biomass Bioenergy,* vol. 15, no. 4-5, pp. 357-367, 1998. [http://dx.doi.org/10.1016/S0961-9534(98)00030-0]

[8] G. Alkhayat, and R. Mehmood, "A review and taxonomy of wind and solar energy forecasting methods based on deep learning", *Energy and AI,* vol. 4, p. 100060, 2021. [http://dx.doi.org/10.1016/j.egyai.2021.100060]

[9] S. Impram, S. Varbak Nese, and B. Oral, "Challenges of renewable energy penetration on power system flexibility: A survey", *Energy Strategy Reviews,* vol. 31, p. 100539, 2020. [http://dx.doi.org/10.1016/j.esr.2020.100539]

[10] I. Ghalehkhondabi, E. Ardjmand, G.R. Weckman, and W.A. Young II, "An overview of energy demand forecasting methods published in 2005–2015", *Energy Systems,* vol. 8, no. 2, pp. 411-447, 2017. [http://dx.doi.org/10.1007/s12667-016-0203-y]

[11] A. Krechowicz, M. Krechowicz, and K. Poczeta, "Machine learning approaches to predict electricity production from renewable energy sources", *Energies,* vol. 15, no. 23, p. 9146, 2022. [http://dx.doi.org/10.3390/en15239146]

[12] Y.Y. Hong, and T.R.A. Satriani, "Day-ahead spatiotemporal wind speed forecasting using robust design-based deep learning neural network", *Energy,* vol. 209, p. 118441, 2020. [http://dx.doi.org/10.1016/j.energy.2020.118441]

[13] X. Zhao, J. Liu, D. Yu, and J. Chang, "One-day-ahead probabilistic wind speed forecast based on optimized numerical weather prediction data", *Energy Convers. Manage.,* vol. 164, pp. 560-569, 2018. [http://dx.doi.org/10.1016/j.enconman.2018.03.030]

[14] H. Wang, C. Ma, and L. Zhou, "A brief review of machine learning and its application", *International Conference on Information Engineering and Computer Science,* pp. 1-4, 2009. [http://dx.doi.org/10.1109/ICIECS.2009.5362936]

[15] K. Liakos, P. Busato, D. Moshou, S. Pearson, and D. Bochtis, "Machine learning in agriculture: A review", *Sensors (Basel),* vol. 18, no. 8, p. 2674, 2018. [http://dx.doi.org/10.3390/s18082674] [PMID: 30110960]

[16] K. Alanne, and S. Sierla, "An overview of machine learning applications for smart buildings", *Sustain Cities Soc.,* vol. 76, p. 103445, 2022. [http://dx.doi.org/10.1016/j.scs.2021.103445]

[17] I.H. Sarker, "Machine learning: Algorithms, real-world applications and research directions", *SN Computer Science,* vol. 2, no. 3, p. 160, 2021. [http://dx.doi.org/10.1007/s42979-021-00592-x] [PMID: 33778771]

[18] N.E. Benti, M.D. Chaka, and A.G. Semie, "Forecasting renewable energy generation with machine learning and deep learning: current advances and future prospects", *Sustainability (Basel),* vol. 15, no. 9, p. 7087, 2023. [http://dx.doi.org/10.3390/su15097087]

[19] S. Xie, and Y. Liu, "Improving supervised learning for meeting summarization using sampling and regression", *Comput. Speech Lang.,* vol. 24, no. 3, pp. 495-514, 2010. [http://dx.doi.org/10.1016/j.csl.2009.04.007]

[20] D. Maulud, and A.M. Abdulazeez, "A review on linear regression comprehensive in machine learning", *Journal of Applied Science and Technology Trends,* vol. 1, no. 2, pp. 140-147, 2020. [http://dx.doi.org/10.38094/jastt1457]

[21] E. Gambhir, R. Jain, A. Gupta, and U. Tomer, "Regression analysis of COVID-19 using machine learning algorithms", *2020 International Conference on Smart Electronics and Communication (ICOSEC),* pp. 65-71, 2020. [http://dx.doi.org/10.1109/ICOSEC49089.2020.9215356]

[22] P. K. Josephine, V. S. Prakash, and K. S. Divya, "Supervised learning algorithms: A comparison," KristuJayanti Journal of Computational Sciences (KJCS), pp. 01-12, 2021.

[23] A.V. Kumar, P. Sivakumar, A. Chaturvedi, I.B. Musirin, V.S. Akula, R.V. Suganya, G. Vanishree, R.H. Pillai, G. Jagadamba, G. Kaur, A. Srinivasulu, and U.N. Dulhare, "Advancements in Metaverse Security: Phishing Website Detection Through Optimal Feature Selection and Random Forest Classifier", In: *Metaverse Security Paradigms.,* B. Gupta, Ed., IGI Global, 2024, pp. 335-374. [Online] [http://dx.doi.org/10.4018/979-8-3693-3824-7.ch014]

[24] W.S. Noble, "What is a support vector machine?", *Nat. Biotechnol.,* vol. 24, no. 12, pp. 1565-1567, 2006.
[http://dx.doi.org/10.1038/nbt1206-1565] [PMID: 17160063]

[25] G. Mountrakis, J. Im, and C. Ogole, "Support vector machines in remote sensing: A review", *ISPRS J. Photogramm. Remote Sens.,* vol. 66, no. 3, pp. 247-259, 2011.
[http://dx.doi.org/10.1016/j.isprsjprs.2010.11.001]

[26] H.U. Dike, Y. Zhou, K.K. Deveerasetty, and Q. Wu, "Unsupervised learning based on artificial neural network: A review", *IEEE International Conference on Cyborg and Bionic Systems (CBS),* pp. 322-327, 2018.
[http://dx.doi.org/10.1109/CBS.2018.8612259]

[27] A. Glielmo, B.E. Husic, A. Rodriguez, C. Clementi, F. Noé, and A. Laio, "Unsupervised learning methods for molecular simulation data", *Chem. Rev.,* vol. 121, no. 16, pp. 9722-9758, 2021.
[http://dx.doi.org/10.1021/acs.chemrev.0c01195] [PMID: 33945269]

[28] J. Karhunen, T. Raiko, and K. Cho, "Unsupervised deep learning: A short review," in Advances in Independent Component Analysis and Learning Machines, pp. 125-142, 2015.
[http://dx.doi.org/10.1016/B978-0-12-802806-3.00007-5]

[29] N. Kannan, and D. Vakeesan, "Solar energy for future world: - A review", *Renew. Sustain. Energy Rev.,* vol. 62, pp. 1092-1105, 2016.
[http://dx.doi.org/10.1016/j.rser.2016.05.022]

[30] S. Mekhilef, R. Saidur, and A. Safari, "A review on solar energy use in industries", *Renew. Sustain. Energy Rev.,* vol. 15, no. 4, pp. 1777-1790, 2011.
[http://dx.doi.org/10.1016/j.rser.2010.12.018]

[31] O. Abedinia, N. Amjady, and N. Ghadimi, "Solar energy forecasting based on hybrid neural network and improved metaheuristic algorithm", *Comput. Intell.,* vol. 34, no. 1, pp. 241-260, 2018.
[http://dx.doi.org/10.1111/coin.12145]

[32] C. Rigollier, O. Bauer, and L. Wald, "On the clear sky model of the ESRA — European Solar Radiation Atlas — with respect to the heliosat method", *Sol. Energy,* vol. 68, no. 1, pp. 33-48, 2000.
[http://dx.doi.org/10.1016/S0038-092X(99)00055-9]

[33] A.T. Degaetano, K.L. Eggleston, and W.W. Knapp, "A comparison of daily solar radiation estimates for the Northeastern United States using the Northeast Regional Climate Center and National Renewable Energy Laboratory models", *Sol. Energy,* vol. 55, no. 3, pp. 185-194, 1995.
[http://dx.doi.org/10.1016/0038-092X(95)00044-R]

[34] D. Yang, E. Wu, and J. Kleissl, "Operational solar forecasting for the real-time market", *Int. J. Forecast.,* vol. 35, no. 4, pp. 1499-1519, 2019.
[http://dx.doi.org/10.1016/j.ijforecast.2019.03.009]

[35] J. Boland, S. Farah, and L. Bai, "Forecasting of wind and solar farm output in the australian national electricity market: A review", *Energies,* vol. 15, no. 1, p. 370, 2022.
[http://dx.doi.org/10.3390/en15010370]

[36] M.K. Singla, J. Gupta, R. Kumar, M. Safaraliev, M.H. Alsharif, A.A. Aly, and M.K. Kim, "A multi-model evaluation of Enhanced Tunicate Swarm Optimization for parameter identification", *Energy Rep.,* vol. 12, pp. 2049-2061, 2024.
[http://dx.doi.org/10.1016/j.egyr.2024.08.015]

[37] M.K. Singla, M. Safaraliev, J. Gupta, M. Aljaidi, I. Odinaev, R. Kumar, and A. Abdel Menaem,

"Optimizing proton exchange membrane fuel cell parameter identification using enhanced hummingbird algorithm", *Int. J. Hydrogen Energy,* vol. 89, pp. 342-352, 2024. [http://dx.doi.org/10.1016/j.ijhydene.2024.09.211]

[38] G. Kaur, M. Shobana, F. Kavin, S. Sellakumar, D. Meenakshi, and N. Bharathiraja, "Secured intelligent transportation with privacy retention through blockchain framework", *J. Intell. Fuzzy Syst.,* vol. 46, no. 4, pp. 10507-10521, 2024. [http://dx.doi.org/10.3233/JIFS-230831]

[39] A.V. Kumar, S.A. Roshan, A. Dutta, S. Ray, S.R. Masadeh, G.P. Lakshmi, D. Michalopoulos, R. Nyayapati, I.B. Musirin, and G. Kaur, "Rainfall Prediction Using Machine Learning", In: *Advancements in Climate and Smart Environment Technology.,* J. Mabrouki, M. Azrour, Eds., IGI Global, 2024, pp. 100-113. [Online] [http://dx.doi.org/10.4018/979-8-3693-3807-0.ch009]

[40] R. Al-Hajj, A. Assi, B. Neji, R. Ghandour, and Z. Al Barakeh, "Transfer learning for renewable energy systems: A survey", *Sustainability (Basel),* vol. 15, no. 11, p. 9131, 2023. [http://dx.doi.org/10.3390/su15119131]

[41] J. Lago, F. De Ridder, and B. De Schutter, "Forecasting spot electricity prices: Deep learning approaches and empirical comparison of traditional algorithms", *Appl. Energy,* vol. 221, pp. 386-405, 2018. [http://dx.doi.org/10.1016/j.apenergy.2018.02.069]

# CHAPTER 6

# Energy Planning and Management: A Holistic Approach

**Pradeep Ramesh Sonar**[1,*], **Sachin Trambak Mahale**[1] and **Ganesh Sakharam Chavan**[2]

[1] *School of Management, D Y Patil University, Pune 410506, Mahahrashtra, India*

[2] *Faculty of Management Studies, Parul University, Vadodara 391760, Gujarat, India*

**Abstract:** This chapter explores the fundamentals of energy planning and management, emphasizing the significance of sustainable practices in today's industrial environment. By introducing the innovative framework, the chapter, "Harmony of Creation: Integrating the 14 Ratnas of Samudra Manthan with the Technological Tapestry of Industry 4.0," explains how traditional values might guide current energy policies by bridging old wisdom with contemporary technology.

The 14 Ratnas of Samudra Manthan serve as metaphors for key elements of energy planning and management. Each Ratna is linked to a principle or technology within Industry 4.0, offering new insights into sustainable practices. For instance, Amrita symbolizes the circular economy, emphasizing strategies focused on resource optimization and waste reduction, which are essential for increasing effectiveness and reducing energy usage. Similarly, Kamadhenu represents resource efficiency, highlighting conservation methodologies that coincide with energy-saving goals.

The chapter explores how to optimize energy management procedures by integrating Machine Learning (ML) and Artificial Intelligence (AI). It provides forecast models for energy use, operational effectiveness, and resource usage using resources such as Monte Carlo simulations. This approach underscores the importance of ethical governance and innovation in developing responsible energy strategies.

By weaving together the ancient wisdom of Samudra Manthan with Industry 4.0 advancements, this chapter provides useful advice for scholars, industry leaders, and policymakers. It highlights the potential for traditional principles to guide the ethical and sustainable use of advanced technologies, contributing to a balanced and responsible future in industrial energy management.

**Keywords:** Artificial intelligence, Energy management, Industry 4.0, Machine learning, Ratnas of Samudra Manthan.

* **Corresponding author Pradeep Ramesh Sonar** School of Management, D Y Patil University, Pune 410506, Mahahrashtra, India; E-mail: drnidhi0927@gmail.com

**Reetu Malhotra, Pooja Mahajan, Manish Singla & Ramesh Kumar (Eds.)**

# OVERVIEW OF ENERGY PLANNING AND MANAGEMENT

"Energy that is not utilized is energy that is conserved." The vivid phrase outlines a basic premise: improving the level of energy management can be achieved through limiting the evident wasting of energy, which leads to more sustainable energy consumption [1]. For instance, rising energy production does not always equate to more efficient energy use. It is a matter of making better use of the assets that we already possess. In the present day and age, wherein many industries are witnessing a radical change, energy planning and management [2] have assumed great significance as they affect the efficiency of the processes [3], sustainability factors, and the level of competition.

## Definition and Importance of Energy Planning in Modern Industries

Energy planning is a methodical process that determines the best ways to produce, distribute, and use energy in order to meet present and future demands with the least amount of expense and environmental harm [4]. Another way to think about energy planning is as a collection of procedures for managing energy demand, resolving political and administrative obstacles to the development of energy, and other energy-related activities. This type of labor encompasses a broad range of tasks, including forecasting future energy consumption, managing supply, ensuring energy security for the advancement of the economy and society, and utilizing renewable energy sources.

Energy [5] is the pulse to all industry activities as it aids in the process of production and is essential in the application of technology and services in industries. Proper energy planning allows enterprises to operate without interruption for extended periods while protecting the environment. Effective energy management [6, 7] helps businesses become more environmentally conscious by lowering expenses and reducing carbon emissions. Energy planning [8] is vital for attaining energy efficiency, cutting waste, and guaranteeing long-term profitability as enterprises transition to sustainable practices [9]. One significant challenge in modern energy management is the need to balance traditional energy sources (such as fossil fuels) with renewable energy sources (such as wind, solar, and hydroelectric power) [10]. The effective integration of both types is essential for achieving sustainability goals [11]. Without strategic planning, industries face the risk of energy shortages, rising costs, and greater environmental impact, as shown in Fig. (**1**).

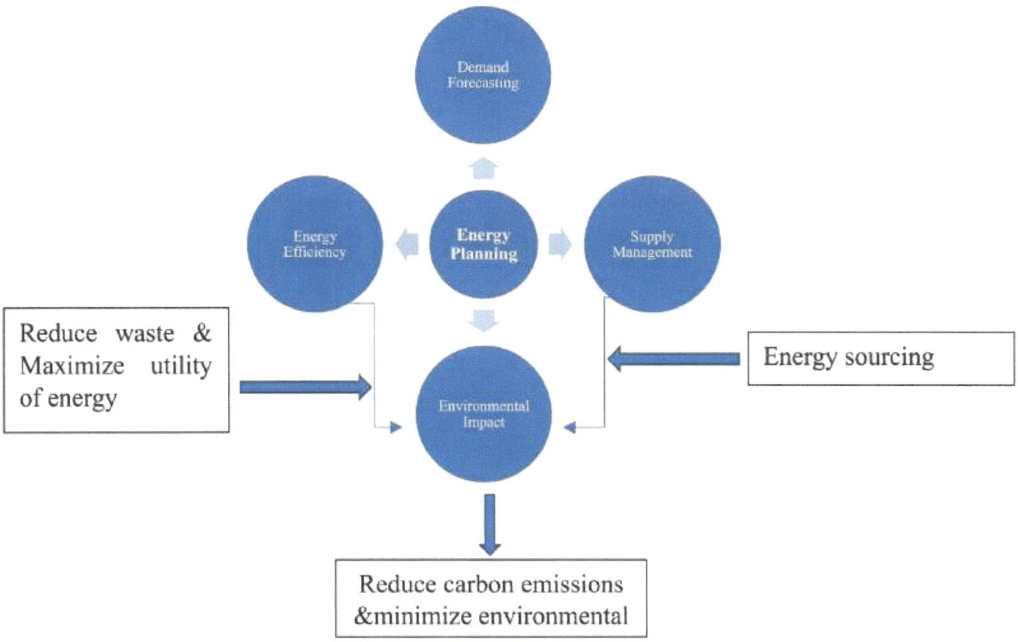


**Fig. (1).** Energy Planning.

Energy planning is at the core of the ecosystem, representing the following aspects.

Demand Forecasting: Predicts future energy needs based on historical data, market trends, and industrial growth.

Supply Management: Involves sourcing energy through a mix of traditional and renewable sources [12].

Energy Efficiency: Focuses on implementing technologies and practices to reduce waste and maximize energy utility [13].

Impact on the Environment: Tries to minimize environmental deterioration and cut down on carbon emissions.

Arrows in the Fig. (**1**) show the relationships and interactions between these components:

Demand Forecasting impacts Supply Management (ensuring that supply meets forecasted demand).

Supply Management influences Energy Efficiency (sourcing strategies affect efficiency practices).

Energy Efficiency affects Environmental Impact (efficient use can reduce environmental harm).

Environmental Impact can feed back into Demand Forecasting (regulations and goals can influence demand predictions).

The above diagram explains how each component of energy planning interacts with and influences the others.

## Emphasis on Sustainability in Energy Management

In the last several years, energy management [14] has made sustainability a major priority. The goal of sustainable energy management is to provide for current energy needs without sacrificing the capacity of future generations to provide for their own needs. Using energy responsibly and efficiently is not a choice but a necessity in this day and age of climate change and depleting natural resources.

Energy management strategies that support international sustainability objectives like the Sustainable Development Goals (SDGs) of the United Nations are becoming more and more popular in industry adoption, as shown in Fig. (**2**) [4]. Energy audits, the application of energy-efficient technologies [15], the incorporation of renewable energy sources, and the adoption of circular economy concepts to minimize waste are some examples of these activities. By emphasizing sustainability, businesses can lessen their reliance on energy and become more resilient to changes in energy prices.

Incorporating sustainability into energy management also brings financial benefits [16, 17]. According to McKinsey & Company, industries that prioritize energy efficiency and sustainability can reduce energy consumption by 20% to 30%, leading to significant cost savings over time. Moreover, many governments offer incentives to industries that invest in renewable energy technologies, further enhancing the financial viability of sustainable energy management.

Economic benefits [18 - 20] include lower operating expenses and cost savings from energy efficiency.

Impact on the environment: less waste, less greenhouse gas emissions, and resource conservation.

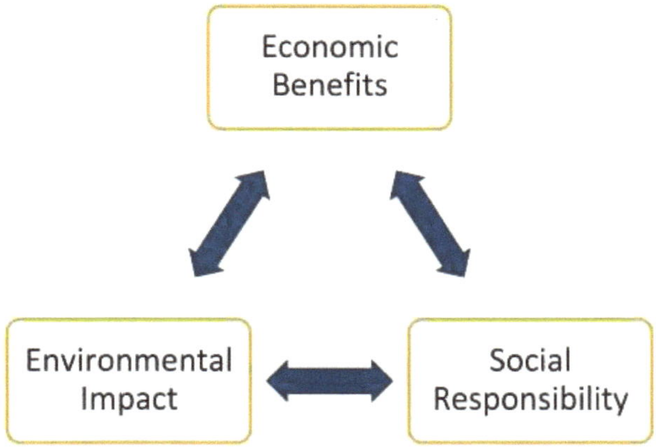


**Fig. (2).** Sustainability Triangle in Energy Management

Social Responsibility: Improving community relations and meeting societal expectations for responsible energy use.

**The Evolving Role of Technology in Shaping Energy Management Strategies**

Technology plays an ever-increasing role in modern energy management. The rise of Industry 4.0, marked by automation, the Internet of Things (IoT), Artificial Intelligence (AI), and big data analytics, has transformed how industries approach energy planning and management [8]. These technologies allow industries to monitor, analyze, and optimize energy usage in real-time, resulting in significant improvements in efficiency and sustainability.

***Smart Energy Management Systems (EMS)***

These systems integrate AI and IoT to monitor energy consumption across facilities, detect inefficiencies, and suggest optimizations. Real-time data gathering on energy consumption is made possible by IoT sensors, and AI algorithms use the data to forecast future energy requirements and pinpoint areas for improvement.

**Predictive Maintenance:** Using IoT sensors, predictive maintenance helps industries maintain equipment efficiency by alerting managers of potential failures before they occur, thereby reducing energy waste. Predictive maintenance, for instance, can stop an energy-guzzling machine by scheduling repairs in a timely manner.

**Renewable Energy Integration:** Renewable energy sources like solar and wind power may now be seamlessly integrated into industrial energy systems thanks to technological advancements. Batteries and other energy storage technologies enable businesses to store extra renewable energy for use at a later time, guaranteeing a steady supply of electricity even in the absence of natural resources.

**Blockchain for Energy Trading**: Blockchain technology is transforming energy markets by enabling decentralized energy trading between producers and consumers. This means that industries generating excess renewable energy can sell it directly to other businesses, creating an efficient and transparent energy market.

**Energy Analytics:** With the help of big data analytics, industries can gain deep insights into their energy usage patterns. This enables them to make data-driven decisions about energy sourcing, load management, and sustainability initiatives.

## Bridging Traditional Wisdom with Modern Energy Management Technologies

The story of Samudra Manthan from ancient Indian scriptures provides a symbolic lesson for energy management in the modern age [15]. Just as the ocean was churned to bring forth valuable resources, industries today must "churn" their energy resources wisely to achieve long-term benefits [5]. The traditional wisdom of balance and efficient resource use aligns with the goals of modern energy management, where energy is not wasted but harnessed to its full potential.

Incorporating traditional wisdom alongside cutting-edge technologies provides a balanced approach to energy planning. Although efficiency and sustainability would seem to be improved by advanced technology, traditional wisdom's ideas of responsibility, balance, and restraint are also employed in resource management and industry to benefit society.

## Industry 4.0 and Its Relevance to Energy Management

Industry 4.0 marks the integration of advanced technologies such as IoT, AI, robotics, and big data analytics into industrial processes. These technologies are revolutionizing energy management by making it more intelligent, flexible, and sustainable. Energy management in Industry 4.0 is data-driven, with real-time monitoring and optimization of energy consumption throughout the supply chain.

With Industry 4.0, industries can achieve energy savings by:

**Automation:** Smart systems can automate energy-intensive processes, ensuring that energy is only used when needed.

**Real-Time Monitoring:** IoT sensors enable continuous monitoring of energy use, allowing industries to identify inefficiencies and take corrective actions instantly.

**Energy Forecasting:** By using previous data, big data analytics can forecast future energy demand, which helps businesses plan their energy use more efficiently.

It is impossible to exaggerate the importance of Industry 4.0 to energy management. Industries that integrate this cutting-edge technology can lower energy usage while simultaneously improving sustainability, cutting expenses, and maintaining their competitiveness in a market that is becoming more and more ecologically concerned.

## CONCEPTUAL FRAMEWORK: THE 14 RATNAS OF SAMUDRA MANTHAN

### Historical and Cultural Context of Samudra Manthan

The ancient Indian story of Samudra Manthan, or the churning of the ocean, is a profound allegory found in the Puranas (ancient scriptures). It represents the quest for Amrita, the nectar of immortality, and the emergence of 14 divine treasures or Ratnas during this epic cosmic event. In this story, both gods (Devas) and demons (Asuras) collaborated to churn the ocean using Mount Mandara as the churning rod and the serpent Vasuki as the rope. This powerful act, despite its challenges, led to the emergence of valuable and destructive forces from the ocean depths, symbolizing both rewards and obstacles inherent in great endeavors.

The tale of Samudra Manthan serves as a metaphor for human efforts to extract value from the chaotic and unpredictable forces of nature. The Ratnas (treasures) that emerged symbolize principles that can guide modern energy management, particularly in the realm of sustainability, resource optimization, and strategic planning.

In today's world, where industries seek to maximize value while minimizing waste, the lessons of Samudra Manthan offer timeless wisdom. Just as the gods and demons worked together despite their differences, modern industries must balance competing needs such as efficiency, profitability, and environmental responsibility to extract the "nectar" of sustainable growth.

## The 14 Ratnas and Their Symbolic Significance in Energy Management

Each of the 14 Ratnas (treasures) holds deep symbolic meaning, representing essential principles in modern energy management. Below is a brief discussion of each Ratna and how it metaphorically connects to modern energy strategies.

**1. Lakshmi (Goddess of Wealth)** – Efficient Resource: Allocation Lakshmi, the goddess of wealth, symbolizes prosperity. In energy management, wealth can be equated with resources-whether they are fossil fuels, renewable energy, or human capital. Just as Lakshmi represents the careful management of wealth to ensure prosperity, industries must allocate energy resources efficiently to ensure both profitability and sustainability. Businesses in Industry 4.0 should aim for shared prosperity, making sure that the advantages of technology improvements are shared fairly among all parties involved, such as the environment, communities, and workers.

For example, using the renewable energy sources of sun and wind effectively would have a long-term financial and environmental payoff.

**2. Kaustubha (A Precious Gem)** – Energy Optimization. The Kaustubha gem, worn by Lord Vishnu, symbolizes clarity of vision and brilliance. In energy management, this can be related to energy optimization—the ability to see inefficiencies clearly and address them. It reflects the need for transparency in energy use, where industries use data analytics and real-time monitoring to optimize energy consumption.

Example: Smart grids and Energy Management Systems (EMS) provide real-time data to help industries optimize energy use, reducing waste.

**3. Parijat (A Celestial Tree)** – Renewable Energy Sources: The Parijat tree is a living example of regeneration or rejuvenation since it blossoms eternally. This highlights the value of self-renewing renewable resources as opposed to fossil fuel resources, which eventually run out.

Example: Solar energy harnesses the infinite power of the sun, much like how the Parijat tree symbolizes endless blooms, contributing to sustainable energy management.

**4. Airavata (A White Elephant)** – Energy Storage: Airavata, the celestial elephant, represents strength and reliability. It represents energy storage devices, such as batteries, which are necessary for dependable and effective energy management in terms of modern energy [17]. In order to balance sporadic renewable energy sources like solar and wind, energy storage is essential.

Example: Lithium-ion batteries store excess energy produced by solar panels during the day, ensuring a continuous energy supply during the night.

**5. Kamadhenu (A Divine Cow)** – Sustainable Energy Practices: Kamadhenu, the wish-fulfilling cow, is a symbol of abundance and nurturing. In the context of energy, it represents sustainable energy practices that nurture the environment. Just as Kamadhenu provides for all needs, sustainable energy management ensures that industries can meet their energy demands without depleting natural resources.

Example: The use of bioenergy and biofuels derived from organic matter, which can regenerate and provide sustainable energy.

**6. Uchhaishravas (A Flying Horse)** – Speed and Innovation: Uchhaishravas, the divine horse, symbolizes speed and innovation. In modern industries, rapid technological innovations in energy management are critical to staying competitive. The speed at which industries adopt new technologies like AI and IoT in energy management determines their efficiency.

Example: AI-driven energy management platforms that can predict energy demand and optimize supply chains in real-time.

Example: Companies like Tesla innovating in the space of electric vehicles and energy storage solutions.

**7. Sharanga (A Bow)** – Strategic Energy Planning: The bow Sharanga, associated with precision and power, represents strategic planning in energy management. Just as the bow is aimed with focus, industries must strategically plan their energy use, targeting efficiency and sustainability goals with precision.

Example: Energy audits that help industries develop targeted strategies for reducing energy consumption.

**8. Panchajanya (A Conch)** – Communication and Collaboration: The conch Panchajanya symbolizes communication. Effective communication and collaboration between different departments (finance, operations, and sustainability) are essential for successful energy management. The Shankha represents awareness and awakening. In energy management, this translates to increasing awareness about the importance of energy conservation and sustainability among employees, stakeholders, and consumers.

Example: Companies running energy conservation campaigns and training programs to promote sustainable energy practices.

Example: Cross-functional teams in industries working together to achieve energy efficiency goals, sharing data and insights from various departments.

**9. Dhanvantari (Physician of the Gods)** – Preventive Maintenance: Dhanvantari, the divine physician, represents healing and prevention. In energy management, this symbolizes preventive maintenance, where regular upkeep of machinery ensures that energy is not wasted due to inefficient equipment.

Example: Predictive maintenance using IoT sensors can alert industries when equipment is about to fail, preventing energy wastage and unplanned downtime.

**10. Varuni (Goddess of Liquor)** – Fluid Energy Flow: Varuni, the goddess of liquor, represents fluidity and flow. This reflects the importance of fluid and flexible energy distribution systems in industries, ensuring a smooth and uninterrupted energy flow. Varuni may also represent responsible consumption. In Industry 4.0, businesses can promote responsible consumption by developing smart products and services that are designed for durability, repairability, and upgradability, reducing the overall environmental impact.

Example: Smart grids that allow for the dynamic distribution of electricity, balancing energy loads based on real-time demand.

**11. Apsara (Celestial Dancer) - Employee Engagement:** Apsara, the celestial dancer, signifies beauty and grace. In Industry 4.0, businesses can prioritize employee engagement, creating a positive and inclusive workplace culture that values diversity, creativity, and collaboration.

**12. Chandra (Moon) - Sustainable Governance:** Chandra, the moon, represents governance and control. In Industry 4.0, sustainable governance involves ethical decision-making, transparent business practices, and adherence to environmental and social regulations.

**13. Amrita (Nectar of Immortality)** – Sustainable Energy Solutions: Amrita, the nectar of immortality, symbolizes sustainable energy solutions that ensure long-term survival and growth. In a future where environmental stewardship is valued more and more, industries that embrace sustainable practices and renewable energy sources will survive.

An illustration of this would be the switch to sustainable energy sources like hydropower, solar power, and wind in order to maintain long-term profitability and abide by environmental laws.

**14. Poison (Halahala)** – Risks and Challenges: The poison Halahala, though dangerous, was necessary to reach the nectar. This symbolizes the risks and challenges that industries must navigate in energy management, such as the initial costs of adopting renewable technologies or the volatility of energy markets.

Example: The high upfront costs of installing solar panels, though challenging, are outweighed by the long-term benefits of lower energy costs and sustainability.

## INDUSTRY 4.0 AND ENERGY MANAGEMENT

### Introduction to Industry 4.0

The term "industry 4.0," or "fourth industrial revolution," describes the present trend in manufacturing and other industries toward automation and data exchange. To build smart factories and streamline procedures, it combines cutting-edge technology, including cloud computing, big data, Artificial Intelligence (AI), the Internet of Things (IoT), and advanced analytics [2]. As a result of these technologies coming together to form Industry 4.0, energy management is changing, guaranteeing increased sustainability, efficiency, and resource usage.

### Key Technological Components of Industry 4.0

The Internet of Things, or IoT, is a network of connected objects and systems that allows for real-time data collecting, monitoring, and energy consumption control.

Artificial Intelligence (AI): Uses machine learning and predictive analytics to forecast energy demand, optimize usage, and automate decision-making processes.

Big Data: Analyzes large volumes of data to uncover patterns, trends, and insights for improving energy efficiency.

Cloud Computing: Provides scalable and accessible storage and processing power for data-driven energy management systems.

Cyber-Physical Systems: Integrate computational processes with physical systems, enabling real-time energy monitoring and adaptive management strategies.

Samudra Manthan's Ratnas and Industry 4.0 in Energy Management in Fig. (**3**).

**Fig. (3).** A painting depicting the various scenes of the Samudra Manthan

Each of the 14 Ratnas (treasures) produced during the churning of the cosmic ocean (Samudra Manthan) can be metaphorically linked to key principles in modern energy management. Industry 4.0 technologies play a crucial role in implementing these principles.

### Lakshmi (Wealth) – Efficient Resource Allocation

Modern Principle: In energy management, resource allocation refers to the judicious use of energy resources to minimize waste and maximize utility.

Industry 4.0 Link: By anticipating energy requirements and modifying usage in real-time, AI and big data enable effective resource allocation and optimize energy use depending on supply and demand.

### Kaustubha (Gem) – Energy Optimization

Modern Principle: Optimizing energy means reducing energy waste and increasing operational efficiency.

Industry 4.0 Link: IoT sensors and data analytics offer comprehensive insights into patterns of energy consumption, facilitating ongoing system and process optimization for increased energy efficiency.

## Parijat (Tree) – Renewable Energy Sources

Modern Principle: The use of renewable energy sources like solar, wind, and hydro power ensures long-term sustainability.

Industry 4.0 Link: Smart grids powered by IoT and AI enable seamless integration of renewable energy into the grid, balancing load and energy storage in real-time.

## Airavata (Elephant) – Energy Storage

Modern Principle: Energy storage systems, such as batteries, are critical for ensuring the stability and reliability of renewable energy sources.

Industry 4.0 Link: AI-driven energy storage management systems can predict peak demands and efficiently manage the discharge and recharge cycles of batteries.

## Kamadhenu (Cow) – Sustainable Energy Practices

Modern Principle: Sustainability focuses on energy conservation and the reduction of carbon footprints [6].

Industry 4.0 Link: Real-time energy tracking, consumption reduction, and integration of sustainable practices into business operations are made possible by IoT and cloud-based technologies.

## Uchhaishravas (Horse) – Speed and Innovation

Modern Principle: Rapid innovations in energy technology ensure that businesses stay competitive.

Industry 4.0 Link: Rapid prototyping and AI-powered innovation systems accelerate the development of new energy-efficient technologies, driving continuous improvement in energy management.

## Sharanga (Bow) – Strategic Energy Planning

Modern Principle: Strategic planning in energy management ensures that long-term goals such as sustainability, cost savings, and regulatory compliance are met.

Industry 4.0 Link: Predictive analytics and AI-based simulations allow companies to strategically plan energy usage, predict future needs, and adjust strategies accordingly to meet goals efficiently.

## Panchajanya (Conch) – Communication and Collaboration

Modern Principle: Achieving energy efficiency requires effective coordination and communication between stakeholders and energy systems.

Industry 4.0 Link: IoT-enabled networks and cloud platforms facilitate real-time data sharing and collaboration between different stakeholders, from energy providers to businesses and consumers.

## Dhanvantari (Physician) – Preventive Maintenance

Modern Principle: Preventive maintenance minimizes downtime and ensures that energy systems run efficiently.

Industry 4.0 Link: Predictive maintenance systems, powered by AI and IoT, monitor equipment health and predict potential failures, reducing unplanned outages and improving energy efficiency.

## Varuni (Liquor) – Fluid Energy Flow

Modern Principle: Maintaining fluidity in energy distribution is essential for a consistent and reliable energy supply.

Industry 4.0 Link: IoT and smart grids enable flexible and adaptable energy distribution, guaranteeing that supply and demand are met without interruption.

## Apsara (Celestial Dancer) - Employee Engagement

Principles for Modern Employee Engagement:

Diversity and Inclusion: Foster a workplace culture that celebrates diversity, ensuring every employee feels valued, respected, and empowered to contribute. Promote policies that embrace different perspectives, backgrounds, and talents.

Creativity and Innovation: Encourage creativity by providing platforms for employees to experiment, innovate, and take calculated risks. Recognize and reward creative contributions that drive business growth and transformation.

Collaboration and Communication: Create open channels for transparent communication, ensuring cross-functional collaboration. Utilize digital tools to enhance real-time interaction and break down silos across the organization.

Work-Life Balance: Prioritize employee well-being through flexible work arrangements, mental health initiatives, and wellness programs. A balanced work-life dynamic leads to greater productivity and employee satisfaction.

Continuous Learning and Development: Invest in ongoing learning opportunities, upskilling, and professional growth to help employees adapt to rapid technological changes. Encourage a culture of lifelong learning to remain competitive.

Recognition and Empowerment: Regularly acknowledge employee efforts and accomplishments. Empower employees by involving them in decision-making processes and granting them ownership of projects.

Digital Employee Experience: Leverage Industry 4.0 technologies such as AI-driven employee platforms, Virtual Reality (VR) for training, and data analytics to personalize and enhance the employee experience.

**Chandra (Moon) - Sustainable Governance**

Modern Principles for Modern Sustainable Governance:

Ethical Decision-Making: Incorporate ethical considerations into all business decisions, ensuring alignment with core values, fairness, and the well-being of stakeholders, including customers, employees, and communities.

Transparency and Accountability: Maintain openness in business operations, disclosing Environmental, Social, and Governance (ESG) impacts. Ensure stakeholders can access accurate, clear, and timely information on the company's practices.

Sustainability Integration: Embed sustainability into corporate strategies, setting long-term goals for reducing carbon footprints, conserving resources, and supporting biodiversity [12]. Prioritize sustainability in the supply chain and product lifecycle.

Compliance with Regulations: Ensure adherence to global, national, and local environmental and social regulations, including emissions standards, labour laws, and corporate governance codes. Regularly audit compliance to mitigate risks.

Stakeholder Engagement: Engage with stakeholders, customers, employees, investors, and the community to understand their needs and expectations. Ensure that governance decisions reflect a balance between profitability and social responsibility.

Risk management and Resilience: Create risk management plans that take technical, social, and environmental disturbances into account. Prepare for hazards associated with sustainability, such as resource scarcity and climate change, to strengthen organizational resilience.

Using Technology to Promote Good Governance: Make use of digital technologies like IoT to assess environmental impact in real time, blockchain to track ESG data transparently, and AI to analyze compliance predictions.

**Amrita (Nectar) – Sustainable Energy Solutions**

Modern Principle: The ultimate objective is to provide dependable, sustainable energy solutions that improve the environment and the economy.

Industry 4.0 Link: Artificial Intelligence (AI)-powered energy management systems reduce energy use, integrate renewable energy sources, and develop long-term sustainable energy solutions.

**Poison (Halahala) – Risks and Challenges**

Modern Principle: Achieving energy sustainability requires controlling the hazards related to energy production and use.

Industry 4.0 Link: AI and IoT-powered risk management systems identify possible energy hazards, such as overconsumption and security flaws, and make sure that risks are adequately addressed.

**The Samudra Manthan and Industry 4.0 in Energy Management**

The Samudra Manthan symbolizes the ongoing churning of ideas, technology, and strategies in modern energy management. Each Ratna from the mythological tale represents a principle that finds its parallel in the technologies and innovations of Industry 4.0. Through this churning of ancient wisdom and modern technologies, businesses can create a sustainable, efficient, and optimized energy ecosystem that balances resources, fosters innovation, and promotes sustainability in a rapidly evolving world.

**Business Applications and Strategic Lessons from Samudra Manthan**

Samudra Manthan's story offers several important corporate strategy lessons, particularly in the area of energy management:

**Collaboration for Mutual Benefit:** Just as gods and demons worked together to churn the ocean, businesses must collaborate with governments, consumers, and other industries to achieve sustainable energy goals.

**Balancing Risks and Rewards:** The emergence of both treasures and poison from the churning represents the balance between risks and rewards. In energy management, businesses must weigh the costs of adopting new technologies against the long-term benefits.

**Innovation through Adversity:** The challenges faced during the churning process led to valuable rewards. Similarly, industries that innovate in the face of energy challenges will emerge stronger and more competitive.

**Sustainability as the Key to Longevity:** The quest for Amrita symbolizes the pursuit of immortality, a goal that can only be achieved through sustainable energy management practices that ensure long-term viability.

With Industry 4.0, the future of energy management is smarter, more efficient, and better equipped to handle the challenges and opportunities of the 21st century [13].

## CIRCULAR ECONOMY AND RESOURCE EFFICIENCY: THE FUNCTION OF AMRITA AND KAMADHENU

The goal of the Circular Economy (CE) is to do away with waste and endless resource usage [7]. On the other hand, the conventional linear economy adheres to the "take-make-dispose" paradigm. The circular economy emphasizes prolonging the life of goods, machinery, and infrastructure in order to increase resource productivity and lessen adverse environmental effects.

This model aligns with the Sanskrit concept of "*Amrita*", the nectar of immortality, which symbolizes sustainability and longevity in resource use. Similar to Amrita's depiction of eternal sustainability, a circular economy maximizes resource usage and minimizes waste to guarantee that energy and resources flow through systems efficiently.

### Amrita and Resource Optimization in the Circular Economy

In Samudra Manthan, Amrita, the nectar of immortality, signifies an endless, sustainable source of life and prosperity. The objectives of a circular economy, in which resources are maintained in an endless cycle of reuse, recycling, and regeneration, are reflected in this concept.

**Waste Reduction:** By prolonging the life cycle of materials and goods, the circular economy seeks to reduce waste, just like amrita, which is an infinite resource. Industry 4.0 technologies—such as IoT sensors, data analytics, and Artificial Intelligence (AI)—can support this process by monitoring and minimizing waste in both production and consumption.

The optimization of resources is a crucial aspect of energy management. Minimizing waste and optimizing energy consumption are in line with the Amrita principle, which stands for using resources to the utmost extent possible. Businesses may anticipate demand, streamline manufacturing procedures, and put real-time energy management plans into place to avoid overproduction and overuse with the aid of big data and artificial intelligence.

**Sustainability:** Ensuring that resources are managed sustainably, with the least amount of negative environmental impact, is at the heart of the circular economy. Amrita, a representation of eternal life, highlights the significance of sustainability in resource management by guaranteeing the availability of resources for future generations. Businesses can incorporate renewable energy sources into their operations and promote long-term sustainability by utilizing cloud-based platforms and smart grids.

## Kamadhenu and Resource Efficiency

In the churning of Samudra Manthan, Kamadhenu, the cow of plenty, is a symbol of abundance and resourcefulness. Kamadhenu's ability to provide endless resources, from milk to other valuable products, can be metaphorically linked to the modern principles of resource efficiency in the circular economy.

### *Resource Efficiency in Energy*

Kamadhenu's symbolism emphasizes the importance of resource efficiency getting more from less. Similarly, resource efficiency in energy management means using energy and resources in ways that minimize waste and maximize productivity.

### *Energy Monitoring and Optimization*

In the circular economy, Industry 4.0 technologies like IoT enable constant monitoring of energy use, helping industries manage resources efficiently [9]. For instance, real-time tracking of usage, inefficiency detection, and energy optimization are all possible with smart energy management systems. This illustrates Kamadhenu's capacity to supply in large quantities while making sure resources are used to their best advantage.

### *Predictive Maintenance*

Predictive maintenance enabled by AI and machine learning guarantees that energy systems function at maximum efficiency, minimizing energy waste and averting needless downtime, just as Kamadhenu continuously provides resources.

By using this technology, companies may ensure optimal resource utilization by addressing possible errors before they occur.

## Role of Industry 4.0 in Enhancing Circular Economy and Resource Efficiency

Industry 4.0 technologies play a key role in advancing both the circular economy and resource efficiency [3]. These technologies integrate data-driven solutions to monitor, analyze, and optimize resource usage across industries.

**IoT Sensors:** IoT sensors are embedded in machinery and infrastructure, tracking real-time resource consumption and energy use. In a circular economy, these sensors can help businesses monitor resource flow and make adjustments to avoid waste.

**AI and Machine Learning:** Algorithms driven by AI can forecast demand, maximize resource allocation, and reduce overproduction or underutilization. Machine Learning (ML) models in resource efficiency allow for system optimization and predictive maintenance to stop energy waste.

**Big Data and Analytics:** Data analytics is useful in a circular economy because it can be used to monitor material lifecycles, forecast resource demand trends, and maximize resource recycling and reuse. Big Data lets firms analyze their consumption patterns and make data-driven decisions, which contributes to resource efficiency.

**3D Printing and Additive Manufacturing:** By enabling businesses to make goods when needed, these technologies help to minimize material waste and overproduction. By guaranteeing that materials are only utilized when necessary, this complies with the circular economy's tenets of resource conservation. An illustration of Amrita and Kamadhenu in relation to resource efficiency and the circular economy may be found below.

As shown in this diagram (Fig. **4**), in the circular economy, Amrita stands for resource optimization and sustainability through the unbroken flow of resources.

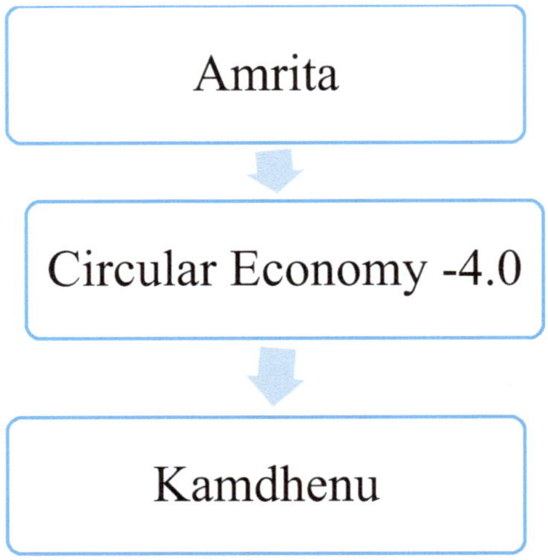


**Fig. (4).** Circular Economy and Resource Efficiency.

Kamadhenu emphasizes the need to reduce waste and optimize resource utility. He stands for the concepts of abundance and resource efficiency.

The circular economy is made possible by Industry 4.0 technologies, which use cloud computing, IoT, AI, and Big Data to improve efficiency and resource optimization. Consequently, sustainability and resource efficiency are important, and this is reflected in the circular economy and modern energy management concepts that incorporate the symbols of Amrita and Kamadhenu. These ideas are strengthened by Industry 4.0 technologies, which offer cutting-edge instruments for resource and energy flow monitoring, optimization, and maintenance. By utilizing data-driven insights, real-time monitoring, and predictive analytics, enterprises may synchronize their processes with the circular economy's objective of optimizing resources, thereby fostering a more sustainable future.

## ARTIFICIAL INTELLIGENCE AND MACHINE LEARNING IN ENERGY MANAGEMENT

By facilitating more effective, anticipatory, and sustainable approaches to energy usage, Artificial Intelligence (AI) and Machine Learning (ML) have revolutionized the energy management industry. Energy management systems incorporate AI and ML technologies to maximize the utilization of renewable resources, minimize waste, and optimize energy consumption [14].

## AI and ML in Energy Management Systems

Energy Demand Forecasting: By utilizing past data, weather trends, and user behavior, AI and ML algorithms are able to predict energy consumption with high accuracy. These algorithms can forecast patterns in energy usage by analyzing massive datasets, which aids utility companies and industries in optimizing supply and demand management.

Machine learning techniques like Long Short-Term Memory (LSTM) neural networks are used extensively for precise prediction of day-ahead and seasonal energy demand patterns. Likewise, Support Vector Machines (SVM) are utilized for anomaly detection in energy consumption, Reinforcement Learning is used for real-time decision optimization in smart grids, and K-Means clustering aids consumer segmentation in demand-side energy management.

**Predictive Maintenance:** Predictive maintenance is one of AI's significant contributions to energy management. Machine learning algorithms have the ability to track machinery and equipment, anticipate malfunctions, and plan maintenance before issues arise, saving energy and preventing downtime.

**Energy Optimization:** AI can optimize energy use in real-time by adjusting systems according to current energy requirements. AI, for instance, may regulate lighting, HVAC (Heating, Ventilation, and Air Conditioning) systems, and other energy-consuming systems in large buildings or smart networks to reduce unnecessary usage of energy.

**Smart Grids**: AI is a critical component of smart grids, which are advanced systems for monitoring and distributing electricity. AI analyzes energy usage patterns, integrates renewable energy sources, and balances supply and demand in real time, making the grid more efficient and reliable.

**Renewable Energy Integration:** The grid can more easily include renewable energy sources like solar and wind power thanks to AI and ML. Artificial Intelligence (AI) is able to dynamically modify energy storage and distribution to guarantee a steady supply by forecasting the availability of renewable resources (such as sunshine and wind speed).

## Case Studies Demonstrating the Impact of Artificial Intelligence and Machine Learning in Energy Management

Google Data Centres: Google has successfully implemented AI to manage its data centres' energy usage. The company applied DeepMind's AI system to optimize cooling systems, which resulted in a 40% reduction in energy use for cooling and a 15% improvement in overall energy efficiency [12].

Siemens' Smart Grid Solutions: Siemens utilizes AI to enhance its smart grid technology, integrating renewable energy while maintaining grid stability. By balancing energy supply and demand, Siemens' AI-driven solutions improve energy distribution's efficiency and dependability [13].

Enel's AI-Powered Energy Management: AI is used by Enel, a multinational energy corporation based in Italy, to control the production of electricity from renewable sources [14]. Artificial Intelligence (AI) algorithms optimize the integration of solar and wind energy into the grid and forecast energy generation based on meteorological conditions, as shown in Fig. (**5**).

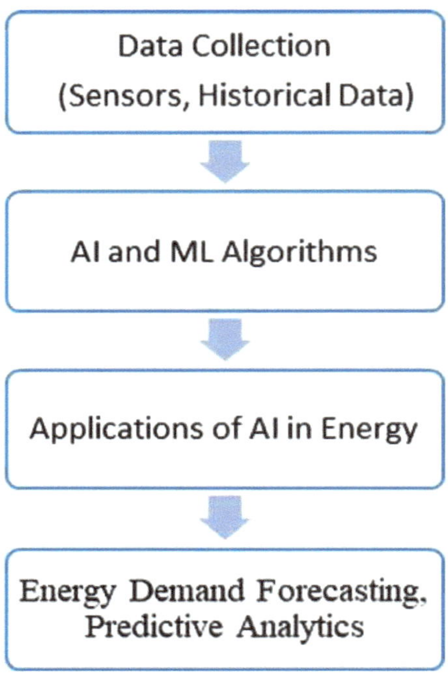


**Fig. (5).** AI and ML in Energy Management.

## ETHICAL GOVERNANCE AND SUSTAINABILITY

The increasing use of AI and other advanced technologies in energy management raises ethical issues that need to be resolved to guarantee the responsible use of these tools. The strategies and concepts that direct the moral development and application of technology while giving sustainability and the well-being of society top priority are referred to as ethical governance.

## Ethical Challenges in AI-Driven Energy Management

Data Privacy: In order for AI systems to work well, enormous volumes of data are required. AI systems gather data from sensors, smart meters, and other devices for use in energy management. But this poses questions regarding data privacy, especially when it comes to sensitive or personal data.

**Algorithmic Bias:** AI systems may exhibit biases that affect decision-making in energy management. For example, algorithms may prioritize energy distribution in areas with higher economic activity, potentially leading to unequal access to energy in underserved regions.

**Accountability and Transparency:** The application of AI to energy management may give rise to issues with transparency and accountability. It may be challenging to assign blame for the outcome if AI-driven judgments have unfavorable effects. Developing transparent AI systems that enable human supervision and accountability is crucial [10].

### Sustainability Initiatives and the Necessity of Responsible Governance

Sustainable Energy Practices: Responsible governance emphasizes the integration of sustainability into energy management practices. AI and ML can help industries transition to sustainable energy sources by optimizing the use of renewables, reducing waste, and improving energy efficiency.

**Environmental Impact:** The ethical governance of AI in energy management necessitates giving significant thought to how it may affect the environment. Although AI can optimize energy utilization, energy is also required for the creation and maintenance of AI systems. Because of this, it is imperative to make sure AI technologies are created and applied in a way that reduces their carbon footprint.

Social Responsibility: Social responsibility is a component of ethical governance. AI integration into energy management should provide fair access to energy and refrain from escalating already-existing disparities. It is imperative to establish policies that guarantee marginalized communities are taken into account during the design and implementation of AI systems.

## CONCLUSION

This chapter discusses the integration of ancient wisdom, specifically the principles from the Samudra Manthan, with modern technological advancements in energy management. The symbolism of the Ratnas provides a metaphorical framework for understanding the critical principles of resource optimization, efficiency, and sustainability in contemporary energy management practices.

## SUMMARY

Traditional Wisdom Meets Modern Technology: Ancient concepts like Amrita and Kamadhenu from the Samudra Manthan have strong parallels in today's energy management landscape. Amrita symbolizes resource optimization, aligning with the circular economy's goal of minimizing waste. Kamadhenu represents resource efficiency, encouraging the effective use of energy and materials.

Industry 4.0: The development of Industry 4.0 technologies, including big data, IoT, AI, and ML, is essential to the advancement of energy management techniques. By enabling real-time optimization, predictive maintenance, and the integration of renewable energy sources, these technologies help to promote more environmentally friendly energy consumption.

AI and ML: The industry has undergone a transformation thanks to the incorporation of AI and ML into energy management systems, which has improved predictive capacities, decreased waste, and increased energy efficiency. Numerous companies have successfully implemented these technologies, as evidenced by case studies from Siemens, Google, and Enel.

Ethical Governance: The chapter also discussed the moral dilemmas that AI and ML in energy management present, such as accountability, algorithmic bias, and data privacy. In order to guarantee that these technologies are applied ethically, with an emphasis on sustainability and fair access to energy, ethical governance is crucial.

## IDEAS AND SUGGESTIONS FOR FURTHER RESEARCH

AI Synergy with Ancient Principles: Future studies can examine how integrating contemporary technologies with ancient philosophical ideas from many civilizations can improve sustainability in energy management.

Sustainability in AI Development: To make sure that AI technologies themselves are in line with sustainability objectives, research can look into measures to lessen the environmental impact of AI development and operation.

Equitable Energy Distribution: More research is required to determine how AI may help achieve equitable energy distribution, especially in marginalized and underprivileged communities.

Implications of Merging Ancient Wisdom with Industry 4.0

Integrating contemporary technologies with conventional concepts, as those embodied by the Ratnas of Samudra Manthan, provides a comprehensive strategy for energy management. Through the application of traditional knowledge to contemporary technical problems, energy systems that are not only optimal and efficient but also sustainable and morally managed can be developed. Industry 4.0 technologies have the power to completely transform energy management and usher in a time where energy is used wisely and responsibly when they are driven by ethical frameworks and age-old wisdom.

## "ROADMAP FOR FRAMEWORK ADOPTION AND POLICY IMPLICATIONS"

Framework Adoption Roadmap: A phased plan for governments/industries to adopt the "14 Ratna–Industry 4.0" integration.

Stage 1: Awareness and Training (institute conventional principles and Industry 4.0 technology).

Stage 2: Pilot Projects (*e.g.*, implement Amrita-Kamadhenu on circular economy projects).

Stage 3: Integration and Scaling (make best practices institutionalized into energy policy).

Policy Recommendations:

Encourage public-private partnership for technologically-enabled energy sustainability.

Incentivize industries implementing AI/ML to enhance resource efficiency.

Integrate conventional sustainability ethics into energy governance systems.

### Think Out of the Box

Monte Carlo Simulation (MCS) is extensively employed to simulate uncertainty and variability in energy systems as a result of varying demand, variability of renewable energy supply (*e.g.*, solar/wind), and market conditions. MCS allows industries to simulate various scenarios and evaluate risks related to strategic energy choices. For instance:

1. Simulation of solar irradiance variability for integration of renewable energy.
2. Evaluation of cost-benefit under uncertainty of shifting from fossil fuels to renewables.
3. Assessment of equipment failure risk in predictive maintenance planning.

Monte Carlo Simulation (MCS) is also a key tool for simulating uncertainties in energy systems. For instance, MCS can mimic thousands of potential weather or demand situations to optimize the integration of intermittent renewable resources, assisting in more robust energy planning.

## REFERENCES

[1] International Energy Agency (IEA), *Energy Management in Industry: A Guide to Energy Efficiency,* 2023.

[2] McKinsey & Company, *Energy Efficiency in Manufacturing: A Global Perspective,* 2023.

[3] World Economic Forum, "Industry 4.0 and Energy Efficiency: Trends and Innovations",

[4] United Nations, *Sustainable Development Goals: Affordable and Clean Energy,* 2023.

[5] Mahabharata and Vishnu Purana, *Samudra Manthan: The Divine Churning of the Ocean.*

[6] United Nations, *Sustainable Energy for All,* 2023.

[7] Ellen MacArthur Foundation, *Circular Economy Overview,* 2015. [Available from: https://www.ellenmacarthurfoundation.org/circular-economy/overview/concept]

[8] World Economic Forum, *Industry 4.0: Shaping the Future of Energy,* 2020. [Available from https://www.weforum.org]

[9] Accenture, *Digital technologies and the circular economy,* 2018.

[10] M. Sandhu, R. Malhotra, and J. Singh, "IoT Enabled - Cloud based Smart Parking System for 5G Service", *Proc. 1st IEEE International Conference on Industrial Electronics: Developments & Applications (ICIDeA),* 2022pp. 202-207 [http://dx.doi.org/10.1109/ICIDeA53933.2022.9970012]

[11] *Design and Construction of Solar Shed.* vol. 2220. , 2020, p. 140061.R. Malhotra, and B. Singh, AIP Conf. Proc. [http://dx.doi.org/10.1063/5.0001157]

[12] DeepMind, *Reducing Google Data Centers' Energy Usage with AI,* 2020. [Available from: https://www.deepmind.com]

[13] D.R. Malhotra, V. Mehta, and N. Neha, "Construction of a Device for Obstacle Detection", *International Journal of Recent Technology and Engineering (IJRTE),* vol. 8, no. 4, pp. 2312-2315, 2019. [http://dx.doi.org/10.35940/ijrte.D8178.118419]

[14] R. Malhotra, A. Jain A, " IoT-based improved mechanical design for elevator cart Applied Intelligence in Human-Computer Interaction" edited by S.Bansal, P.C.Sharma, A. Sharma, J.R.Chang, (Eds.). 2023. 1st ed. CRC Press. [http://dx.doi.org/10.1201/9781003415466]

[15] S. Arora, M. Aljaidi, M. Sandhu, S.H.A. Khan, and R. Malhotra, "Determination of yearly energy savings by CVR in combination with photovoltaic inverter control functions in distribution networks", *Energy Reports,* pp. 5507-5520, 2024. [http://dx.doi.org/10.1016/j.egyr.2024.11.001]

[16] H. Kaur, and R. Malhotra, *Stochastic Analysis of a Two-Unit Standby Autoclave System with Inspection and Varying Demand.* J. Mechanics of Continua and Mathematical Sciences, 2024, pp. 147-163.

[17] Wikimedia Commons, *File: Sagar mathan (corrected version).* [Available from: https://commons.wikimedia.org/wiki/ File:Sagar_mathan_(corrected_version).jpg]

[18] R. Malhotra, and A. Jain, *"IoT-based Improved Mechanical Design for Elevator Cart," in Applied Intelligence in Human-Computer Interaction.,* S. Bansal, P.C. Sharma, A. Sharma, J-R. Chang, Eds., 1st ed. CRC Press, 2023. [http://dx.doi.org/10.1201/9781003415466]

[19] R. Malhotra, *"Reliability Evaluation with Variation in Demand," Systems Reliability Engineering - Modeling and Performance Improvement.* Gruyter: DE, 2022. [http://dx.doi.org/10.1515/9783110617375]

[20] R. Malhotra, F. S. Alamri, and H. A. E.-W. Khalifa, *Novel Analysis between Two-Unit Hot and Cold Standby Redundant Systems with Varied Demand,* 2023. [http://dx.doi.org/10.3390/sym15061220]

# CHAPTER 7

# Improved Opportunities in Energy Efficiency by using Supply Chain Management

**Sachin Trambak Mahale**[1,*], **Pradeep Ramesh Sonar**[1], **Ganesh Sakharam Chavan**[2] and **Reetu Malhotra**[3]

[1] *School of Management, D Y Patil University, Pune 410507, Maharashtra, India*

[2] *Faculty of Management Studies, Parul University, Vadodara 391760, Gujarat, India*

[3] *Department of Applied Sciences, Chitkara Institute of Engineering & Technology, Chitkara University, Rajpura 140401, Punjab, India*

**Abstract:** Implementing energy efficiency into Supply Chain Management (SCM) is crucial for tackling the issues presented by the global energy deficits. As energy efficiency becomes increasingly essential for economic and social progress, it emphasizes the role of sustainable supply chains in enhancing business competitiveness, profitability, and overall performance. The discussion focuses on the complexities and risks inherent in global supply chains, identifying strategies to improve energy efficiency, particularly for Small and Medium-sized Enterprises (SMEs), which often face barriers such as limited access to capital and a lack of awareness. A comprehensive review of current literature highlights the need for a shift from single-firm energy efficiency to a broader, supply chain-wide approach. By reinforcing key stages of SCM and implementing sustainable practices, companies can achieve strategic goals, reduce costs, mitigate risks, and optimize processes. These offer a valuable framework for the energy sector and other industries, demonstrating how improved supply chain management can significantly enhance energy efficiency and contribute to the sustainability of global operations.

**Keywords:** Energy efficiency, SMEs, Supply chain management, Sustainability operations.

## INTRODUCTION

In the contemporary world, severe energy crises have taken over due to high industrialization [1 - 4], urbanization, population, and the effects of climate change. The energy demand increases in conjunction with environmental problems worldwide; therefore, the situation is complex, and timely action is necessary for sustainable development.

* **Corresponding author Sachin Trambak Mahale** School of Management, D Y Patil University, Pune 410507, Maharashtra, India; E-mail: sachin.mahale@dypatiluniversitypune.edu.in

**Reetu Malhotra, Pooja Mahajan, Manish Singla & Ramesh Kumar (Eds.)**

It is estimated that global energy demand will grow by about 25% by 2040, primarily due to emerging economies such as China and India, as well as other high-growth markets. Industrial usage accounts for approximately 42% of the total global energy consumption. China alone consumes nearly 30 percent of industrial energy. In other parts of the world, in Southeast Asia, Sub-Saharan Africa, and Latin America, there is economic growth, which increases energy use in industries, transport, and households, increasing pressure on global energy resources.

It complicates the energy crisis because of the further rapid expansion of urbanization and population growth. In 2022, the world had a population of 8 billion people, and it is estimated that by 2050, the world population will increase to 9.7 billion, with most of this growth in urban regions. Urban regions account for 75% of the world's energy consumption, thereby placing greater pressure on energy supplies. An increasingly large share of the world's energy demand, primarily in cities, is fueled by fossil fuels, which conflicts with efforts to mitigate climate change.

The largest emitter of greenhouse gases globally remains the source of energy production, which accounts for 72% of all emissions. According to the Intergovernmental Panel on Climate Change, the combustion of fossil fuels and resource forms such as oil, gas, and coal drives a cocktail of toxic pollutants into the air, causing serious damage to the environment and dimming the world's ability to reach the International Climate Change Goals set by the Paris Agreement. Examples include rising world temperature to less than 1.5 degrees Celsius above pre-industrial levels.

IRENA believes that enhancing energy efficiency might reduce global energy consumption by up to 40% by 2050, making it a very effective strategy for addressing the issue. Ultimately, the implementation of energy-efficient technology in industrial, transportation, and domestic applications reduces operational expenses and fosters environmental sustainability.

Improving energy efficiency is a significant part of managing any supply chain. It would lead to optimization in the logistics, transportation, and even production processes. This can be achieved by a reduction in fuel consumption: smart logistics can save up to 15% of fuel usage. Big multinational companies like Walmart, Apple, and Unilever have been adopting green practices in their supply chains while trying to reduce their carbon footprint. Transition toward renewable energy models and circular economy models is necessary to enhance energy efficiency and achieve environmental goals.

Energy efficiency is the key to sustainable development, as it encompasses the causes of environmental, economic, and social concerns. According to the United Nations Environment Programme, "sustainable development means that meeting the needs of the present without compromising the ability of future generations to meet their own needs." All the aspects that come under this concept include energy efficiency, reduced consumption, as well as resource conservation and minimizing their negative impacts, which in turn help to stimulate growth.

Energy efficiency is good not only for the environment but also in relation to a highly critical element of sustainable development, namely energy security. Improved energy use can reduce the high consumption of imported fossil fuels. This can be conceived such that a 1% increase in global energy efficiency could lead to a 2% decrease in demand for imported energy. Such reduced utilization further enhances the national energy security, stabilizes energy prices, and boosts the economies, especially in places with critical infrastructures.

Energy efficiency is now becoming a crucial change inducer for renewable energy sources and sustainable development. Overall, this demand can be met by the smaller share from renewable sources such as solar, wind, and hydro power sources. According to the agency, by combining energy efficiency with renewable energy technologies, carbon reductions will account for more than 90% of the required decrease to attain the global climate goal by 2050.

Energy efficiency creates an innovation culture with high productivity and competitiveness of the firms. Therefore, organizations embracing energy efficiency have fewer operating costs than their counterparts. They can, hence, use the saved resources for innovations in research and development [5]. The report by the World Economic Forum concludes that industries implement energy efficiency practices to support supply chains and build their competitive capacity. However, these energy-efficient technologies have also introduced new markets, ranging from retrofitting and developing new concepts, and thereby providing new employment and sustainable economic growth.

It is a crucial element in solving the international energy crisis and delivering the agenda of sustainable development. Energy efficiency cuts across both greenhouse gas emissions and energy security, as well as the opportunity to shift to renewables, integrating many aspects into supply chains and other critical sectors of the global economy. This approach helps mitigate the challenges posed by the international energy crisis and ensures long-term prosperity.

## HISTORICAL PERSPECTIVES ON ENERGY EFFICIENCY AND SUPPLY CHAIN MANAGEMENT

Energy efficiency and Supply Chain Management (SCM) date back to the Industrial Revolution [6]. During the Industrial Revolution, energy resources were tapped to mass produce goods. Energy was consumed in inefficient manners since, at that moment, output was primary, instead of using the amount of energy. As industries grew, so did the importance of energy costs; thus energy-efficient practices began to be more visibly needed.

Such issues were countered through the development of more efficient machine technology and regulations by the 20th century. Among these, the most significant milestone was the U.S. Environmental Protection Agency's Energy Star program, launched in 1992. Under the program, Energy Star supported the adoption of energy-efficient products and practices, influencing more industries to make decisions that make their business operations more energy-conscious; it later became the benchmark for energy efficiency improvement at the international level [7].

Concurrently with this, the emergence of supply chain management as a leading practice has occurred during the intensification of globalization. Increased needs to reduce operational costs among businesses led to an integral supply chain that not only became efficient in the flows of goods and services but also focused on minimizing energy consumption. Energy efficiency becomes a vital strategy that would help produce a competitive advantage, aligned with sustainability objectives.

With the advancement of organizations into the 21st century, the culture of energy efficiency became more and more embedded with supply chain management due to corporate social responsibility and the attention that needed to be paid to environmental considerations. This has led many to realize the potential for significant savings in terms of money, reduced carbon emissions, and operational efficiencies due to energy management within a supply chain network [8]. This historical development reflects growing sentiment that the energy efficiency incorporated into SCM brings both environmental and economic benefits.

## DEFINITION OF ENERGY EFFICIENCY IN SCM

In SCM, energy efficiency refers to the maximization of energy use at all supply chain stages, including sourcing, procurement, production, and logistics, while minimizing waste and maximizing productivity. Organizations can reduce operational expenses, minimize the release of greenhouse gases, and improve sustainability by integrating energy-efficient processes and technologies. Energy efficiency in SCM not only enhances the environmental performance of organizations but also supports their economic objectives by cutting energy expenses and increasing output per energy unit used [9].

The use of intelligent logistics systems for optimizing routes in transport can minimize fuel consumption, whereas renewable sources during the production process will reduce non-renewable resources. Mentzer *et al.* (2001) defined energy efficiency as being vital for making the supply chain more resilient and sustainable in the long term, particularly with the growing trend of regulatory pressure on enterprises to reduce their carbon footprint." Another important point that Christopher suggests is about energy-efficient SCM, enhancing operational efficiency and CSR, driving innovation and competitive forces in present markets [10].

## Benefits of Integrating Energy Efficiency into SCM

The integration of energy efficiency into SCM provides significant environmental and economic benefits. Energy-efficient SCM enhances business resilience, reduces costs, and ensures sustainability in the supply chain. According to Chopra and Meindl (2016), the optimization of energy use at any stage of transportation, manufacturing, or warehousing has resulted in lower operational costs [6]. A logistics process integrated with energy-efficient technologies is known to reduce fuel consumption and energy costs with increased profitability.

1. Cost Optimization: Energy-efficiency activities, such as efficient logistics and renewable energy, optimize the cost of operations, thus offering more profit percentages [11]. Energy-saving production mechanisms tend to lower energy costs.
2. Sustainability & Reduced Emissions: Energy consumption is reduced, and it leads to reduced greenhouse gas emissions, which means that companies would now comply with current environmental regulations and conform with global sustainability goals like the Paris Agreement.
3. Energy Security: Companies relying on renewable energy sources minimize dependency on fossil fuels, making them secure from the volatility and unavailability of the price of energy.
4. Corporate Image & Compliance: Companies using energy-saving means enhance their corporate image while complying with regulatory requirements. This increases consumer as well as investor confidence and saves from non-compliance penalties on environmental regulations at the same time.
5. Operational Efficiency & Innovation: Energy efficiency stimulates innovation as it employs the applications of technologies like IoT and AI that help save

energy, hence increasing productivity.

6. Competitive Advantage: Those firms with energy-efficient supply chains are ready to respond to market trends and regulatory changes well, thus giving them a competitive advantage in being in sync with consumer preferences about green goods.

Thus, an energy-efficient SCM cuts costs, supports sustainability, enhances energy security, and strengthens a corporate image in the market.

**Case Studies: Successful Integration of Energy Efficiency in SCM**

The bottom line: Several organizations have successfully implemented energy efficiency into the management of a supply chain, which has been proven to result not only in cost savings but also reductions in pollution. Examples of some of the organizations that have managed to enhance their supply chains through the use of energy-efficient techniques are provided below.

***Walmart: Energy-Efficient Logistics and Green Sourcing***

Walmart has made huge investments in ways to optimize logistics and sourcing activities. Route optimization, biodiesel energy-efficient trucks, and alternative fuels have made Walmart to cause a reduction in fuel consumption; as a result, the company now avoids emitting over 650,000 metric tons of $CO_2$ each year (EDF). In addition, through its Sustainability Index, it collaborates with suppliers to improve energy efficiency and achieve zero emissions by 2040.

***Unilever: Renewable Energy Integration in Manufacturing***

Such is the case of Unilever, which is a multinational consumer goods company that has made efforts to integrate renewable energy sources in its production process [12]. Indeed, by 2020, a number of countries were completely utilizing renewable energy sources in their manufacturing process. In fact, it achieved this feat by acquiring solar and wind power, making the company eliminate more than 1 metric ton of $CO_2$ every year. Moreover, it also retrofitted its factories with new energy-efficient lighting systems and HVAC systems, reducing energy consumption by 20% globally.

***Toyota: Lean Manufacturing and Energy Efficiency***

Toyota's production system, known as the Toyota Production System (TPS), with waste reductions, now encompasses efficiency in energy. Using smart energy management systems, the Toyota plants track and adjust their energy usage in real-time, which reduces waste. Toyota aims to reduce $CO_2$ emissions by 30% by 2025 through the adoption of energy-efficient robotics, heat recycling, and promoting environmentally responsible purchasing and supply chains in its automotive production process.

### *Nestlé: Energy Efficiency in Transportation and Cold Chain*

The company has been promoting energy efficiency in its cold chain logistics network, ensuring the efficient transport of temperature-sensitive products. Nestlé reduced 50% energy usage in refrigerated transportation through energy-efficient refrigeration and alternative cooling. The company is also utilizing electric and hybrid vehicles to reduce fuel consumption as it strives for carbon neutrality by 2050, as outlined in its Net Zero Roadmap.

### *Apple: Carbon-Neutral Supply Chain*

Apple aims to make its entire supply chain carbon neutral by 2030. To achieve this challenge, the firm has partnered with over 70 suppliers to switch to renewable energy sources. Efficiency in Apple's own supply chain, through high-efficiency equipment, has reduced energy use by 20 percent at some of the company's plants. In one year, 2021, this avoided the release of more than 13 million metric tons of $CO_2$.

These case studies show how the infusion of energy efficiency into the management of supply chains enables corporations to reduce costs, meet sustainability and corporate social responsibility goals. By embracing such innovative energy practices, companies are drastically cutting down on their impacts on the environment while improving operational efficiency.

## CHALLENGES FACED BY SMES IN IMPLEMENTING ENERGY EFFICIENCY

Small and medium-sized enterprises play a huge role in international supply chains, but with regard to energy efficiency measures, they are uniquely challenged. Due to the dilemmas posed by the lack of access to resources, meager technical know-how, obsolescence of infrastructure, and financial constraints, it is challenging for SMEs to adopt sustainable strategies.

### *Limited Financial Resources*

Such much-needed capital as initial capital to finance energy-efficient technologies and infrastructure is not available, since the majority of these investments call for high up-front investments. Renewable energy systems or energy-efficient machinery require financial investment that most small-scale enterprises cannot afford. According to Trianni *et al*. (2013), SMEs tend to find energy efficiency less appealing compared to short-term operational needs, mainly due to their small profit margins and limited access to financing, which hinders most SMEs from taking an active role in energy-saving initiatives [13].

### *Lack of Technical Knowledge and Expertise*

Specialized knowledge is often lacking in energy efficiency, which is likely to be a challenge for SMEs. High technologies, for example, involve energy management systems and smart manufacturing, which require specialties that SMEs cannot afford. Energy performance evaluation, in many SMEs, lacks staff, and the same applies to energy-saving strategy implementation. Even more critical is the cost of hiring external consultants, which exacerbates the gap, preventing the SME from adopting energy-efficient practices [13].

### *Inadequate Infrastructure and Outdated Equipment*

The majority of SMEs use obsolete equipment and infrastructure that cannot be used with state-of-the-art, energy-efficient technologies. For instance, the HVAC systems or smart energy management upgrades require immense retrofitting work, which may bring so much costs for SMEs. According to Rohdin *et al*. (2007), the obsolescence of machinery and processes raises the cost and difficulty of implementating energy efficiency, rendering it unattractive to less large businesses [14].

However, a lower likelihood of stricter regulation against environmental issues, especially against SMEs when compared to bigger organizations, means lower incentive for SMEs in implementing energy efficiency. In the absence of high regulatory and market pressure, SMEs will focus on short-term needs, like customer demand, rather than long-term sustainability. For instance, in competitive markets, Cagno and Trianni (2012) explain that the absence of consumer demand for sustainable products suggests that there is no strong business case for energy efficiency in SMEs.

### *Perceived Complexity and Uncertainty*

Smaller and medium-sized enterprises usually treat energy efficiency projects as complex and uncertain because ROI is not assured. Changes in the core business processes of the company, uncertainty about financial benefits, and changes in energy prices over time make energy efficiency projects appear risky. SMEs are much more risk-averse than other large corporations. As such, they are reluctant to invest in energy efficiency technologies because of unclear and delayed financial returns [15].

### *Limited Access to Energy Efficiency Incentives and Programs*

Though the intentions of most governments and organizations are very good, offering incentives to make energy efficiency common, SMEs generally face bureaucratic or unawareness-related or application complexity barriers in obtaining them. Actual knowledge of the available technologies and programs often dawns late on SMEs, or funds are scarce to assist in this process, making it all the more difficult for SMEs to become energy efficient by availing the offered incentives.

## COMMON BARRIERS TO ENERGY EFFICIENCY IN SMES

There are several barriers that SMEs encounter in implementing measures of energy efficiency, largely due to financial constraints and lack of technical competence, and external pressures.

### *Financial Constraints*

SMEs often lack the capital to invest in energy efficiency, such as upgrading equipment, renewable sources of energy, or energy management systems. These upgrades are costly at the outset, so tight budgets and limited access to funds for SMEs make short-term needs more important than long-term energy efficiency investments. Trianni *et al*. (2013) state that the prime motivation behind why SMEs avoid energy-efficient solutions is based on financial barriers, and the factor of limited awareness regarding government incentives also contributes to the same.

### *Lack of Awareness and Expertise*

Most SMEs have no knowledge or technical know-how in appraising energy efficiency opportunities and their associated savings. Energy management still remains an ancillary function to most businesses, and most SMEs are not staffed with specialized personnel dedicated to energy decisions. A lack of awareness of available energy-efficient technologies prevents SMEs from appreciating the value of such technologies; therefore, it becomes challenging to properly identify and implement cost-effective energy efficiency measures [16].

### *Limitation in Technology Access*

Outdated machinery and infrastructure are the other major challenges facing SMEs. The retrofitting of existing systems to include new energy-efficient technologies can be more expensive and technically complicated. According to Rohdin *et al.* (2007), the smaller companies, especially those in industries that use bulky equipment, face significant hurdles in acquiring the newest, most energy-efficient machinery, mainly because of the high cost incurred and the technical adjustments needed.

### *Short-Term Orientation and Uncertainty of Payback*

SMEs focus predominantly on survival in the short term, and given the uncertain ROI for energy efficiency, they are unlikely to invest. The payback periods of financings for energy-efficient technologies take much more time to recoup in financial benefits, and high front-end expenses are likely financially unsustainable when energy is volatile. Trianni and Cagno argue that a long payback period is a strong disincentive for SMEs as they seek quicker, more predictable paybacks.

### *Lack of Regulatory Pressure*

SMEs usually receive less regulatory pressure on environmental performance than large corporations, hence lowering the urgency level to implement energy efficiency practices. Without the necessary outside pressures, energy efficiency becomes of lesser importance. Hasanbeigi *et al.* (2010) point out that for sectors with fewer environmental regulations, SMEs are least motivated to initiate sustainable practices.

### *Implementation Complexity*

Energy efficiency projects are perceived as labor- and capital-intensive, demanding a change to existing operations, retraining personnel, or shutting down production for retrofits. Such perceived disruption tips the scales on the less-desirable side for energy-saving initiatives and causes SMEs to remain static in implementing energy-saving measures. Complexity in implementing a project, along with the lack of capacity for change, tends to be a barrier for smaller businesses to adapt to energy-efficient technologies.

### *External Support*

Availability of various types of incentives that motivate towards energy efficiency has made it an uphill task for SMEs, since most of such schemes are either bogged down by bureaucratic red tape or general ignorance. Rohdin *et al*. (2007) have also presented that one of the significant hindrances faced by SMEs is that they don't find or access the best available support programs, mainly due to the fact that such programs are less highlighted or their application packages usually accommodate larger businesses in that region [17].

## 6.GLOBAL SUPPLY CHAIN COMPLEXITIES AND RISKS

Global supply chains are complex, especially due to globalization, advances in technology, and shifts in markets. This complexity, therefore, poses a risk to business operations, especially to companies that implement and work towards energy-efficient practices.

Geographical and cultural diversity complicates global supply chains as businesses have to operate according to different regulatory, labor, and environmental settings in the various regions. Miscommunication, along with cultural differences between global suppliers yet further hinders coordination (Christopher, 2016).

In addition to inherent reasons, external factors such as natural catastrophes, political instability, and pandemics make disruptions more common within global supply chains. The COVID-19 crisis brought into limelight the vulnerable scenario of lockdowns, which entail a halt in production and shortages in supplies [18]. Ultimately, these disruptions can lead to inefficiencies and increased energy use in supply chains.

Extended Lead Times and transportation problems only add more lead time to the acquisition of inputs from distant suppliers. This makes scheduling production and logistics complex. There will be a problem of unreliable fuel prices and port congestion that will mitigate the firm's productivity.

Fragmented Supply Chains arise from sourcing inputs from various suppliers, making it difficult to maintain quality, sustainability, and energy standards. This fragmentation only adds to the risk of inefficiencies and supply chain opacity.

Risks in emerging markets comprise infrastructural unreliability, political instability, and inconsistent regulatory frameworks. Emerging markets may offer cost benefits, but they are also associated with much higher risks that may outweigh the sustainability of the supply chain.

Sustainability Pressures from consumers and governments for businesses to increase greener technologies and reduce carbon emissions also complicate global supply chains [19].

Finally, Technological Integration has the potential to bring efficiency-generating effects but requires coordination and investment, particularly in less developed regions [20].

## Variability in Regulations

The variations in other countries' regulations make global supply chains considerably troublesome when discussing labor standards, trade policies, environmental regulations, and energy efficiency. Inconsistencies can make compliance a challenge, drive up costs, and even potentially impede sustainability efforts.

Differences in regulations on environment and energy vary significantly, ranging from strict carbon emission constraints of the EU to no policies in developing countries. It has posed a problem to companies whose supply chains are spread over various geographical locations.

Trade and Tariff Variations also add complexity because companies find themselves adjusting supplier networks or even shifting bases of operations in response to changes in trade policies or tariffs. This will further complicate efforts in implementing energy efficiency. Baldwin and Venables (2013) indicate how the alteration of agreement on trade costs spreads the cost structures and logistics, that is, an actually tangible level of sustainability activities.

Labor and health, along with safety standards, vary profoundly across countries; therefore, they present both ethical and operational risks. Locke *et al.* rightly depict the issue companies face in terms of enforcing uniform standards of labor in supply chains, where performance ultimately has an indirect impact on energy efficiency due to working conditions [21].

Energy pricing and subsidy policies impact the rate of adoption of energy-efficient technologies. Lin and Jiang (2011) argue that areas where energy charges are comparatively lower have little to no incentive towards being energy efficient, thus explaining inconsistencies in energy-saving efforts.

Compliance costs run high because businesses are expected to meet different regulatory standards, meaning higher operations costs and a more complicated uniform implementation of energy efficiency measures.

Lastly, barriers to innovation arise from obsolete or restrictive regulations that inhibit the adoption of energy-efficient technologies. Structured regulatory frameworks may spur or hamper innovation [22].

In a nutshell and generalizing, regulatory variability adds complexity and costs, slowing the journey toward global energy efficiency and sustainability.

### *Market Volatility*

Market volatility affects energy efficiency, financial stability, and operational resilience as the major challenges for global supply chains. Some of the key areas of volatility include: the conditions of raw materials and energy in terms of price volatility, the level of demand, currency exchange rates, geopolitical risks, financial instability, and supply disruptions.

Price Volatility Inflation of raw materials and energy could bring about volatility, thereby disrupting costs in production and logistics costs. Operational cost is directly associated with geopolitical tensions and supply shortages, which result in volatility in the energy market. High-energy-price shocks put pressure on the budget while further delaying investment in energy-efficient technology [23].

The uncertainty in demand variability arises because changes in the preferences of consumers or the economy result in variations in production. According to Lee (2002), the "The bullwhip effect" results in inefficient inventory management and wastes resource utilization, while increasing the energy consumption of supply chains [24].

This is because exchange rate volatility affects the price of importing countries' goods and services, to some extent, determining supply chain investments, and energy efficiency. Ghosh and Constantinides further underpin the fact that sudden changes in exchange rates have caused more detriments in energy-saving initiatives due to increased operational costs.

Geopolitical risks, trade wars, or political instability may cause occasional trade shocks. For example, geopolitical tensions that undermine energy-efficient practices because of higher costs and the swift adjustment in the strategy of the supply chain by organizations have been noted by Bown (2020).

Finally, supply and transportation disruptions, under the scenarios of natural disasters or pandemics, force companies to resort to less energy-efficient alternatives. Indeed, such disruptions may result in chronic impacts on energy efficiency in that companies will react with quick fixes, which are not sustainable, like air freight over sea transport, as postulated by Ivanov and Dolgui (2020).

Market volatility will, therefore, require adaptable, resilient strategies in operations but also on energy efficiency for companies during uncertain times.

## RECENT TRENDS AND STRATEGIES

New technologies, education, and environmental concerns sound familiar in a transformation of business landscapes into something different. The technology of artificial intelligence changes everything in real space. Brynjolfsson and McAfee (2014) state that AI is driving productivity by automating routine tasks, thereby making more time available for strategic and creative problem-solving. AI can also reshape current systems that would make decisions by analyzing a huge amount of data through machine learning. Big data merged with AI algorithms generates actionable insight helpful in the decision-making processes of the organizations. The extent to which organizations depend on data-driven decision-making has increased, but increased dependence on AI has significantly boosted the operational efficiency and strategic planning in most organizations [25].

Simultaneously, the global pandemic accelerated the shift towards online learning platforms on the education side. Dhawan further points out that through online education, people have democratized access to learning, allowing for learning in adaptable and accessible ways, benefiting students worldwide [26]. This online technological tool has helped to change the way teachers can offer customized, adaptive, and better learning results using Learning Management Systems. Virtual reality and augmented reality are other emerging technologies that further explore the educational environment. Mikropoulos and Natsis (2011) correctly point to the fact that the context provided by VR and AR expands in-depth learning as a promising area for future educational systems to increase student focus and understanding [27].

The fourth major area is environmental sustainability, wherein tremendous growth has recently been experienced, including an increasing transformation in systems toward renewable energy sources—wind and solar—sustained and strengthened by progress in technology and supporting regulations. As pointed out by Markard and Rosenbloom (2020), this transition is inevitable in the adoption of alternative sources of energy, other than fossil fuels and in reducing carbon emissions. Besides, the theory of circular economy is on the rise in sustainable development together with the revolution in energy from renewable sources. The circular economy addresses the issues of minimizing waste, conserving resources, and reusing materials, ultimately leading to a sustainable industrial system that respects the environment [28].

Fast-paced changes in AI, the development of online school systems, and the global call for environmental sustainability, among other things, are shaping what is coming around in various industries. With the help of AI, various businesses can enhance productivity and make more informed decisions. Technological advancements in education have proven to be effective in improving learning experiences and accessibility. The key to handling environmental issues will also be a switch to renewable energy and embracing principles of the circular economy. Continuous innovation in such areas promotes economic growth, the progress of society, and the protection of the environment.

## STRATEGIES FOR IMPROVING ENERGY EFFICIENCY IN SUPPLY CHAINS

Improving energy efficiency in a supply chain can reduce costs, promote sustainability, and fulfill some regulatory requirements. There are some effective strategies – from leveraging technological progress to encouraging partnerships among supply chain members.

1. ***Energy Efficiency Technologies***: Advanced automation and analytics-enabled data can make all the difference in the improvement of the entire supply chain process. IoT devices are used to provide direct current insights into energy usage; thus, firms can immediately discover potential inefficiencies and, accordingly, make appropriate corrections. Smart technologies, according to the International Energy Agency, may save 20% on energy for supply chain activities. These kinds of technologies can improve the process efficiency with increased visibility to allow for better business decisions.
2. ***Lean Manufacturing Principles***: In lean principles, the natural tendency is towards waste minimization and continuous improvement. Analysis of all processes can identify areas that have no scarcity of energy for waste, such as transportation and inventory. Just-in-time inventory systems reduce the energy waste of the storage and transportation processes. According to Feldmann & Duflou (2020), the energies in the respective applications can be enhanced between 15% to 25% through lean practices. Promoting an energy-conscious culture is another attribute to be acquired by the employees.
3. ***Supply chain partnerships:*** Such supply chain partners may collaborate to achieve significant improvement in energy efficiency. Platforms like GSCM are more focused on environmental sustainability and therefore also enhance

energy performance [29]. A firm can establish mutual targets with suppliers and logistics service providers regarding improvements in their energy efficiency to achieve substantial energy savings. Consumer support for the programs of sustainability might also be effective.

4. ***Renewable Energy and Sustainable Practices:*** This can be achieved through switching to renewable sources of energy like solar and wind, which can reduce fossil fuel consumption and the emission of GHGs. As McKinsey & Company has suggested, "renewables can help achieve up to 30% energy cost savings in the long term". Furthermore, "green logistics" would also involve proper routes for better transportation and less energy-wasteful, more efficient vehicles, which might lead to energy savings in transport chains and a reduced carbon footprint in supply chains.
5. ***Permanent Monitoring and Reporting:*** Continuous improvement demands permanent monitoring and reporting of energy performance metrics. Determination of the KPIs relevant to energy efficiency will enable an organization to follow up on its progress and decide to improve in those areas. Data obtained from the energy management systems can lead to decisions that ensure that the measures adopted for efficiency factors will align with the business goals. Third-party certifications, such as ISO 50001 can assure the customer regarding the commitment of a company towards energy efficiency and support improvement.

The improvement of energy efficiency in supply chains integrates advanced technologies; adopts the lean principle, collaborates with partners; invests in renewable energy; and monitors performance continuously. All these bring together cost savings for businesses, sustainability benefits, and competitive advantagesin the marketplace.

## SUPPLY CHAIN CONFIGURATION AND DESIGN

Supply chain configuration and design encompasses the strategic laying out of supply chain networks, as well as their components; this includes the structure, processes, and resources needed for the efficient movement of goods, services, and information from suppliers to customers. Properly configured supply chains increase energy efficiency, reduce costs, improve flexibility, and enhance sustainability. Some of the key elements involve network topology, facility locations, transportation modes, inventory management, and supplier relationships. Here is how well-thought-out supply chains can also affect energy efficiency and performance in operations:

## Network Topology and Facility Location

Part of the supply chain design involves optimizing the network topology, including determining the optimal locations for the manufacturing plant, distribution centre, warehouses, and retail outlets. Strategically locating these facilities reduces the distance of transportation considerably, resulting in a significant reduction in energy use. For instance, locating distribution centres near end customers would reduce the time consumed in travel and fuel consumption, hence improving energy efficiency and lead times. Chopra and Meindl (2016) observe that the strategic location of facilities reduces logistics costs and eliminates harmful environmental impacts from transportation, thereby making supply chains more energy-efficient. Tools like the Centre of Gravity method or linear programming are useful to structure supply chains with balanced cost-efficiency and energy-saving features.

## Transportation and Distribution Network

Transportation mode selections are another important configuration aspect of supply chains. There are differences in energy efficiency between different transport modes: road, rail, sea, and air transport modes. Rail transport is a kind of high-energy efficiency when transport distances are big compared to road or air transport. Businesses should be multimodal to maximize energy savings and leverage their strong modes. According to the International Energy Agency, shifting freight from road to rail can reduce transport energy consumption by up to 80% [30]. Cross-docking strategies, where products are transferred directly from the inbound to the outbound vehicle without entering a warehouse, reduce the storage time of goods in warehouses and minimize energy-consuming operations, such as heating, cooling, and lighting, inside these warehouses.

## Inventory Management and Warehouse

Efficient management of inventory is crucial to reducing energy consumption in the warehouse. Inefficient inventory requires significantly more space for storage; it consumes more lighting, cooling, or heating, thereby leading to increased energy consumption. Lean management or efficient inventory management keeps just amount of inventory, thus conserving enough energy. Smart warehouse design also reduces energy consumption by utilizing smart lighting, AS/RS, and energy-efficient HVAC systems.

### Collaboration with Suppliers and Partners

Building a supply chain design also involves identifying the right suppliers and building strong partner relationships. Energy efficiency can be enhanced by cooperating with suppliers and logistics providers. For example, suppliers located farther away geographically, or those aligned with renewable energy best practices, can also reduce the energy footprint of the supply chain. Companies that collaborate in designing the supply chain- companies which share information, synchronize operations, and optimize processes among partners are more energy-efficient, claims the World Economic Forum (2020). Suppliers who use sustainable materials and process management may reduce the energy consumption of their supply chain by a significant amount.

### Technology Integration

The integration of advanced technologies in design may involve the use of IoT, AI, and advanced analytics in the supply chain to monitor and optimize energy usage more effectively. It becomes possible to track the real-time consumption of energy at a number of stages of the supply chain by using IoT devices, whereas AI-based tools may find the best route and production schedule for logistics in minimizing their consumption of energy. Indeed, McKinsey & Company (2021) points out that the integration of smart technology in supply chains' design reduces energy consumption by up to 30%, through optimized decisions and automated procedures that save energy.

A better supply chain design would thus include effective operations; less energy consumption will be achieved due to an optimized topological network, strategic transport modes, efficient stock management, collaboration with suppliers, and an integration of advanced technologies. All these factors come together for better sustainability, reduced costs, and lower energy consumption, as shown in Fig. (**1**).

## COLLABORATION AMONG SUPPLY CHAIN PARTNERS

Collaboration between supply chain partners is crucial for achieving more efficient, cost-effective, and sustainable operations. The global economy is increasingly interconnected, meaning that no single firm can operate in isolation. Improved resource allocation, waste reduction, and process efficiency are certainly valuable benefits that can be achieved through enhanced collaboration with higher-value suppliers, manufacturers, logistics providers, and retailers. The rest of it explores the role that coordination plays in improving energy efficiency and overall supply chain performance.

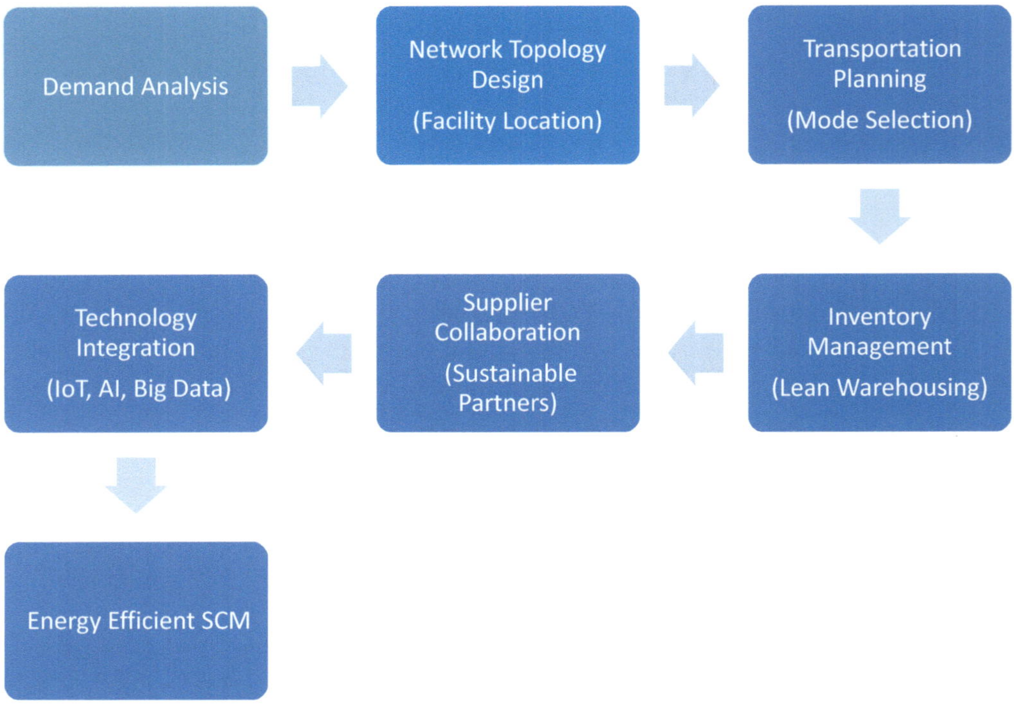


**Fig. (1).** A Key Element for Supply Chain Configuration and Design.

## Information Sharing and Transparency

Supply chain collaboration thrives on information sharing. Open data sharing in terms of inventory, demand forecasts, production schedules, and logistics plans between partners may help harmonize operations, reduce energy consumption, and efficiently use resources. Transparency will pinpoint the inefficiencies and create joint solutions to address these issues. The World Economic Forum (2020) further declares: "Open information sharing reduces lead times, prevents stockouts, and minimizes excess inventory.". This helps reduce energy consumption since fewer expedited shipments, overproduction, and excessive warehousing are needed. CPFR is one of the ways through which this can be provided by having all participants who are involved aligned to ensure that there is adequate data for making appropriate decisions.

## Common Investment in Technology

The best benefit of collaborative endeavors is often common investments that enhance supply chain performance. Partners can also invest in IoT sensors, in conjunction with real-time tracking systems that provide information on location and the condition of goods. This technology also enhances effective inventory management and optimizes routes, saving energy as well. According to McKinsey & Company, shared digital platforms for real-time visibility can reduce energy efficiency up to 20 percent in logistic and warehousing services. These technologies are assisting better coordination in partner activities, thereby ensuring complete truckloads, optimizing routes to be transported, and managing energy-intensive activities much more efficiently.

### Supplier Integration and Co-Innovation

Supplier integration is considered an essential aspect of cooperation. Closer coordination with suppliers enables a company's co-innovation and the development of greener technologies and sustainable practices. Examples may include renewable energy usage in manufacturing or just developing energy-efficient packaging. According to the Harvard Business Review, co-innovation may lead to breakthroughs in sustainability, such as energy-efficient production processes or supply chain systems that minimize environmental impacts. Working with suppliers as partners helps advocate mutual responsibility and expectation for meeting sustainability goals within the supply chain.

### Aligning Goals of Sustainability

Sustainability now ranks center stage, and its goal alignment in supply chains becomes more crucial. In this respect, companies are now teaming up with suppliers and logistics providers to achieve collective sustainability objectives that entail reducing environmental footprints through initiatives such as sustainability targets. Some others include clean energy vehicles or wind power in their warehouses. For the GRI, corporate alignment of sustainability objectives lowers carbon footprints and energy use, coupled with the overall environmental footprint. Collaboration also aids firms in complying with stringent regulations on environmental factors, thereby avoiding fines and enhancing their corporate reputation.

### Risk Sharing and Joint Problem Solving

Shared partnership helps in risk management and joint problem-solving. In the event of delivery failure or a surge in demand, companies may partner to find solutions. Companies partnering with logistics partners found new routes or suppliers during the lockdown period due to the COVID-19 pandemic. Shared risks and shared responsibilities prevent one party from solely bearing the impact of a problem, creating stability and robustness. Shared contingency planning can reduce expensive last-minute adjustments, such as expedited shipping, which usually increases consumption of energy.

Joint collaboration activities among supply chain partners enhance efficiency and sustainability through the sharing of information, joint technology investments, supplier integration, shared sustainability goals, and risk sharing. Collaboration between businesses helps them optimize their operations, reduce the usage of energy, and hence achieve a more sustainable supply chain shown in Fig. (**2**).

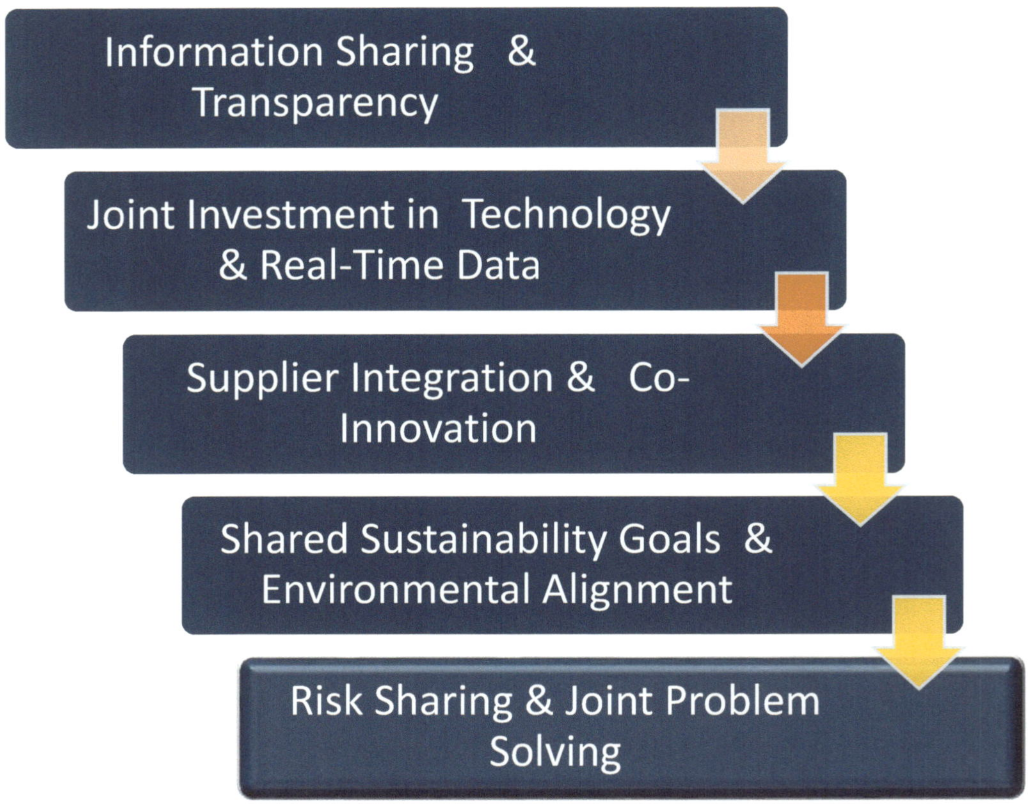


**Fig. (2).** Collaborative steps among Supply Chain Partners.

## *IMPLEMENTING GREEN TECHNOLOGIES*

Supply chains must integrate green technologies in order to enhance environmental sustainability, energy, and reduce carbon footprint. Green technologies, also known as clean or green technologies, involve solutions that minimize the consumption of energy, reduce the emission of greenhouse gases, and promote the use of renewable resources. Technologies addressing these issues encompass areas such as manufacturing processes, transportation systems, packaging solutions, and energy management. Implementation of green technologies will not only abate environmental problems but also reduce the cost over time and enhance the efficiency of operations.

## Renewable Energy Integration

The integration of renewable energy sources from a supply chain, like solar energy, wind energy, and hydroelectric power, is an amazing piece of green technology. For instance, a company can install photovoltaic panels on the rooftops of the warehouses to produce on-site clean energy to power the lighting, heating, or cooling systems, thereby saving a lot of consumption of energy and cost. The International Renewable Energy Agency (IRENA, 2020) reveals that the use of renewable energy lowers carbon emissions by up to 40% over time, especially with the integration of energy storage solutions [31]. The Power Purchase Agreement enables companies to procure renewable energy from a source located outside the enterprise and then reduce their carbon footprint.

## Energy Efficient Manufacturing

Green technologies in manufacturing have to be driven by reduced energy usage and waste. Properly designed energy-efficient machines and smart manufacturing systems, equipped with sensors and automation, can significantly reduce energy consumption. McKinsey & Company (2021) believes that combining smart manufacturing technologies with energy-efficient machines can reduce energy use by as much as 30% with the same or better production output. Apart from this, the incorporation of 3D printing and the use of robots will also eliminate material waste, hence reducing the energy input sourced from the raw material extraction and processing stage.

## Sustainable Transportation Solutions

Transportation is likely to be one of the primary sectors where green technologies will play a significant role in making a substantial difference in terms of energy efficiency. The delivery fleet can be electrified with EVs, and long-haul transport can be reduced by means of hybrid trucks. In this regard, as it is stated from the International Energy Agency (IEA, 2021), electrifying logistics will help reduce energy consumption by up to 60 to 70 percent in comparison to similar conventional diesel- or gasoline-powered vehicles. Route optimization software can also be used, thereby reducing trips and idle time.

## Sustainable Packaging

Green technologies also comprise sustainable packaging. Waste plastic and Styrofoam have become the major polluters and are consuming more energy. Companies have already adopted an alternative material that is easier to biodegrade and recycle, rather than consuming more energy and requiring more effort to dispose of it. The Ellen MacArthur Foundation (2019) revealed that solutions with sustainable packaging options, such as plant-based plastics and minimalist designs, can reduce the carbon footprint in packaging by up to 50%.

## Wastes and Waste Management/Waste Recycling

Waste management and recycling are integral components of green supply chains. The closed-loop system reduces the amount of assimilated and reused waste products. Closed-loop systems reuse or recycle waste products for use, reducing the need for raw material extraction and the energy required for waste disposal. For example, waste-to-energy technologies convert organic waste into energy for use in manufacturing facilities or transportation fleets. According to the World Economic Forum (2020) closed-loop supply chains are potentially capable of saving up to 25% energy especially through the avoidance of virgin materials and waste sent to landfills.

## Smart Energy Management Systems

Smart energy management systems help in efficient usage of energy by leveraging IoT sensors, automation, and advanced analytics to monitor and optimize energy usage. The system can automatically control lighting and HVAC systems based on occupancy and ambient conditions. According to Gartner (2020), the use of smart energy management systems can save 20-30% in overall energy consumption, resulting in notable cost savings and improved environmental performance.

Green technologies encompass integrating renewable energy into the production system, implementing energy-efficient manufacturing practices, promoting sustainable transportation, utilizing sustainable packaging, effective waste management, and utilizing smart energy systems. The implementation of any of these technological advances significantly increases energy efficiency and sustainability in supply chains (Fig. **3**).

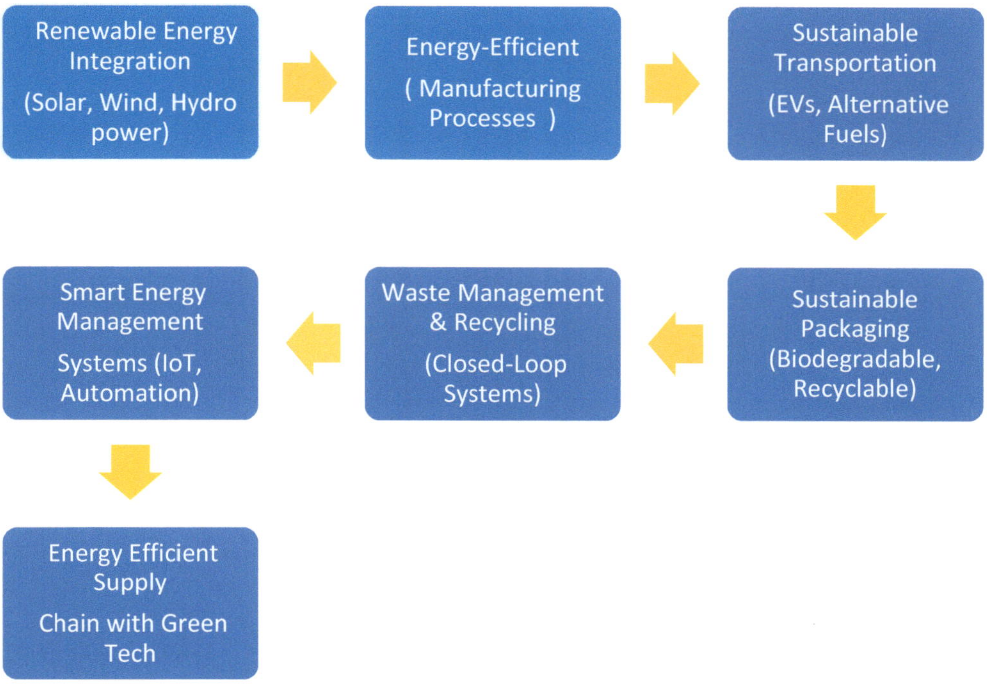


**Fig. (3).** A step-by-step guide for implementing green technologies.

## *Education and Training Initiatives*

Education and training programs play a crucial role in achieving sustainability and energy efficiency in supply chain management. However, as the technologies and approaches in supply chain management evolve and interact, there is a growing need for well-trained personnel with the know-how of operation-based and environmental impacts. The described education and training programs can strengthen employees' knowledge and skills toward sustainable practices, the use of green technology, and decisions that fit in with the objectives of energy efficiency.

### Skill Development

Technologies like the Internet of Things (IoT), Artificial Intelligence (AI), and big data analytics have changed supply chain operations; hence, a skilled workforce is required to exploit these changes. Training should be on the use of IoT for real-time monitoring of energy levels, AI for predictive maintenance, and data analytics to optimize transportation routes and inventory levels. Companies that invest in training for digital and sustainable technologies report significant results concerning operational efficiency and energy savings, according to Deloitte (2020).

## Sustainable Practices in the Supply Chain

All employees at all tiers of the business must be educated on the importance of sustainable supply chain practices, starting with an understanding of the topic of sustainability and energy efficiency through waste minimization, reduction of carbon footprint, and the implementation of eco-friendly practices in procurement, production, and logistics. According to Harvard Business Review (2021), such education trains employees to realize long-term benefits, including cost savings and regulatory compliance, and reinforces corporate reputation. The discussion topics may include the principles of the circular economy, waste reduction, and the use of renewable energy.

## Leadership Training

Leadership training is crucial for developing an energy efficiency and sustainability culture. Programs targeting executives and managers should focus on strategic decisions based on sustainability and energy management. Out of such understanding comes the strategic value that energy-efficient supply chains will bring, which will make it possible for efficient resource allocation, green technology investment, and policy development that will provide fertile grounds for sustainability. The World Economic Forum (2020) indicated that companies focused on sustainability, under appropriate leadership, perform better in terms of both financial performance and energy efficiency than theicounterparts. The training program should apply sustainability frameworks, such as the Triple Bottom Line (TBL), and include strategies for continual improvement in energy usage.

## Certification Programs

Certification initiatives may enhance education and training endeavours through industry-recognized best practice credentials. Examples of these include the Certified Supply Chain Professional (CSCP) and LEED (Leadership in Energy and Environmental Design). Through such certifications, employees' knowledge regarding sustainable supply chain practices is acknowledged. Companies that have enhanced sustainability certifications reap advantages from a better, enlightened workforce with the capability to install energy-saving technologies and practices. Such certification encompasses topics such as sustainable procurement, energy management, and the implementation of green technology, according to Gartner (2021).

### Continuous Learning and Improvement

Continuous learning is encouraged because green technologies and supply chains will always remain changing. Continuous education through workshops, webinars, and online courses ensures that employees are up-to-date on the latest changes in sustainability trends, energy-efficient technologies, and regulations. According to McKinsey & Company (2021), in order for the supply chain to evolve and develop innovation, it must be encouraged through continuous education. Continuity of learning implies seeking new solution approaches and remaining competitive compared to other supply chains that are efficient in their energy usage.

This encompasses education and skills improvement in the workforce through adequate training, sustainable practices, leadership development, certification, and continuous learning—all in pursuit of making strides toward improved energy efficiency and sustainability in supply chains.

## METRICS AND MONITORING FOR ENERGY PERFORMANCE

Metrics and monitoring are absolutely important elements in tracking and improving energy performance in supply chains. Companies can clearly understand their energy consumption patterns, identify inefficiencies, and develop strategies for improving energy efficiency by using the right Key Performance Indicators (KPIs) and monitoring tools. Metrics also serve as a basis for decisions and reporting, which enable organizations to meet regulatory requirements and sustainability targets. This section outlines common elements that appear in energy performance metrics, the value of measurement, and how these practices benefit energy-efficient supply chains.

### Some Key KPIs

Energy efficiency in any supply chain is best analyzed using Key Performance Indicators. Smart metrics help a company monitor the energy used in transportation, warehousing, and manufacturing. The most commonly used KPIs to measure energy performance are:

- Energy Intensity: Measures the amount of energy used to get one unit of output, such as kilowatt-hours per unit produced. A lower energy intensity means better energy efficiency.
- Tons of carbon per unit of product: it is the total amount of greenhouse gases emitted in relation to one unit of product. This provides a better understanding of the environmental implications associated with any activity in the supply chain.
- Energy cost per unit of product: this accounts for the total amount of energy needed to generate one unit of product. If the energy costs per unit decrease, this may be an indication that energy usage has improved.
- Sector-wise energy usage: this keeps track of energy consumption in the sub-activities or sector aspects of the supply chain, such as manufacture, transport, and warehousing, so that companies can identify where they may increase their improvement.

More than 50% of companies, according to Accenture's 2021 report, have clear KPIs for energy performance tracking and will enhance their activities in a continuous loop over the next five years, resulting in a reduction of energy consumption by as much as 25%.

**2. Real-Time Energy Monitoring Systems** Companies are increasingly using real-time energy monitoring systems to track and manage the performance of energy usage. These systems work on IoT sensors, automation, and cloud-based platforms to repeatedly monitor energy usage from every angle and phase of the supply chain. For example, an HVAC sensor within a warehouse can be monitored constantly to track the number of hours it is in use, and a transportation fleet sensor will be able to track the quantity of fuel used as well as the idling hours on the vehicles.

Firms can identify inefficiencies immediately and make quick adjustments to reduce energy consumption with the use of monitoring systems. Predictive analytics can also be focused on estimating future energy consumption based on trends and actual data generated from the company. According to McKinsey & Company (2020), real-time energy monitoring can yield potential savings of up to 15% in the first year as organizations can actively control their energy consumption by preventing unnecessary waste.

### Energy Audits and Benchmarking

Energy performance is tracked through energy audits, which provide insight into how energy is used in supply chain operations and offer opportunities to identify efficiency improvement opportunities. Frequently, these assessments are made regularly and give rise to the development of strategies aimed at saving energy.

Benchmarking is a practice in which a company compares its performance with the standard or that of competitive firms in an industry. This allows companies to evaluate their levels of energy efficiency and, hence, disclose areas for improvement. For example, the U.S. Environmental Protection Agency (EPA) offers the Energy Star Program, which provides benchmarks to numerous sectors, enabling firms to compare the amount of energy they use relative to best-in-class standards.

A study by the International Energy Agency (2021) reveals that companies that usually conduct energy audits and benchmark their operations save between 10-20% in various industries and sizes of operation.

### Data Analytics and Reporting Tools

The application of data analytics in monitoring energy use is becoming increasingly popular. Through the use of these advanced data analytics tools, organizations are now in a position to analyze vast amounts of information that would typically be obtained from sources such as IoT devices, energy bills, and operational records. Organizations can then draw inferences that guide their decisions based on emerging patterns in energy consumption. These advanced tools also enable accurate forecasting, supporting ongoing improvements in energy performance.

Reporting tools can also be used to track trends and report energy performance results to key stakeholders, including investors, customers, and regulators. Increasingly, companies are turning to the use of Global Reporting Initiative standards to report on their energy usage and sustainability efforts, thereby leading to greater transparency and accountability.

According to Deloitte (2020), companies using data analytics and reporting technologies can improve energy efficiency, comply with legal restrictions, and enhance their performance in sustainability. This, in turn, leads to a better reputation in the corporate world, increased customer loyalty, and greater commitment.

### Continual Improvement through Feedback Loops

This is an essential element in good performance monitoring of energy: the composition of feedback loops, which ensures that regularly examined data gathered by monitoring systems feed back into decisions related to energy usage. For example, real-time data from the monitoring system can show spikes in energy usage at specific times of the day, which can help adjust production schedules to avoid these peaks in energy consumption.

However, Feedback Loops enable companies to frequently check their energy performance, newly set goals, and energy-saving measures. According to a 2021 Harvard Business Review study, energy performance with continuous feedback loops leads to operations with better efficiency and higher levels of long-term sustainability compared to a firm that does not implement such feedback loops.

## ESTABLISHING STANDARDS AND BEST PRACTICES

Standardization and best practice for energy efficiency will encourage sustainable supply chains. In an increasing complex and global environment, consistency, transparency, and visible movement will be mandated as standard requirements. Setting standards and practices helps companies reduce their energy usage, while working toward promoting improved energy efficiency throughout their supply chain network. This unifies and aligns all parties involved-from the supplier to the consumer-with common goals of sustainability.

### Standards of Energy Efficiency

Energy consumption in a supply chain is measured through energy efficiency standards. These standards are developed by regulatory bodies, industry associations, or collaborative initiatives and provide measurable guidelines for reporting on energy usage as well as sustainability objectives. Standards and facilitate performance comparison, enable

For instance, the International Organization for Standardization developed the ISO 50001 Energy Management Standard. It is an international framework for establishing energy management systems, which guides a firm in developing energy policies, setting targets, and monitoring progress towards improved energy efficiency. According to a study by Gartner (2021), organizations that adopt ISO 50001 energy management can reduce their energy consumption levels by as much as 20% within five years, thanks to the systematic approach to managing energy. By offering numerous benefits, in addition to the sustainability-enforcing character, such compliance increasingly increases its value in the eyes of investors, customers, and governments.

### Industry-Specific Standards and Collaborative Platforms

Most industries also have sector-specific guidelines and collaborative platforms that promote efficiency in energy consumption, other than global standards like ISO 50001. In this way, they may be better equipped to respond to unique sector challenges.

The Responsible Business Alliance (RBA) has developed specific energy management standards for the electronics industry, focusing on minimizing energy consumption and emissions. According to a report issued by Accenture in 2020, electronics companies that have adopted RBA standards have taken significant steps to save energy through process optimization and reduce energy intensity.

Similarly, the Sustainable Apparel Coalition developed the Higg Index, an index in the apparel industry designed to track environmental and energy impacts. The Higg Index lends itself to transparency and accountability; therefore, it makes it easier for companies to track their energy use, set reduction targets, and share best practices. The World Economic Forum's 2022 statistics assert that companies adopting the Higg Index have been able to lower energy consumption by up to 10% per year by enhancing energy efficiency in their supply chains.

## Best Practices on Supply Chain Energy Efficiency

Best practices on energy efficiency are actually practical methods for optimizing energy use. The best practices of industry cooperation, research and early adoption are frequently adopted through lean manufacturing as well as by employing energy-efficient technologies.

Lean manufacturing, which focuses on minimizing waste and streamlining the entire production chain, is widely adopted in the pursuit of improved energy efficiency. A report from McKinsey & Company, published in 2021, indicates that companies operating with lean practices consume significantly less energy. Specifically, they can reduce energy consumption by up to 30% in cases where non-added production steps, logistics optimization, and improved stock management are implemented.

Another area that could potentially save a significant amount on operational energy usage is through an investment in energy-efficient technologies, such as LED lights, sophisticated HVAC systems, and efficient machinery, among others. For instance, General Electric (GE) collaborated with its suppliers to ensure that all its manufacturing processes are extremely energy efficient. This has significantly reduced energy consumption throughout the supply chain.

## Role of Governments and Industry Associations

Governments, along with industry associations, have to be an extremely important for setting and enforcing energy efficiency standards. While governments frequently impose regulations, requiring companies to meet or exceed certain energy performance criteria, industry associations often provide more voluntary guidelines and certifications that have been developed to enhance standards.

The European Union's Energy Efficiency Directive requires companies operating in its member states to conduct energy audits and implement energy efficiency measures. The Carbon Trust accredits a company if it has achieved any of the listed energy efficiency standards for this accreditation. This is one of the steps that help enhance the level of energy efficiency in a company. PwC (2022) states that the involved companies both conserve energy while at the same time showing allegiance to sustainability, hence achieving brand reputation and competitiveness in the market.

### Continuous Improvement and Monitoring

Continuous improvement and monitoring are essential after standards and best practices have been implemented; companies must maintain continuous improvement by regularly monitoring and updating their strategies. Continuous improvement is a process of setting measurable goals, monitoring progress, and revising practices according to performance data and the emergence of new technologies.

There should be data analytics and energy management software with real-time monitoring to capture energy performance. These software tools provide companies with actionable insights into consumption patterns, enabling them to identify inefficiencies and rectify them. According to Deloitte (2021), companies with advanced energy analytics have improved their energy efficiency in operations by 15% within five years by promptly identifying and resolving inefficiencies in their supply chains.

This can be achieved through the development of standards and practices for sustainable supply chains, focusing on energy efficiency, to be implemented. Companies can improve their supply chain by optimizing energy usage and following that framework as a demonstration of their sustainability concerns.

## POLICY AND REGULATION IN SUPPORTING SUPPLY CHAIN EFFICIENCY

Policy and regulation will play a strategic role in enhancing supply chain efficiency, especially in energy management and sustainability. Government institutions and regulatory bodies create the rules and framework that guide business operations and measures towards energy use efficiency, ensuring the utmost utilization of available resources. Among their other impacts, these policies not only set specific standards but also offer incentives along with guidance on how companies can adopt more efficient processes in all stages of the supply chain.

## Mandatory Standards and Goals

The consequence of energy policies with mandatory standards and targets is that they directly enforce efficiency in the supply chain by requiring companies to meet predetermined thresholds of energy performance. Those regulations typically require businesses to adopt energy-efficient technologies, become more efficient, and closely monitor their energy usage.

For example, the European Union's Energy Efficiency Directive (EED) obliges the states to carry out measures for the improvement of energy efficiency in industries like manufacturing and transportation. In return, it provides binding targets on reduced energy consumption, along with systematic energy audits. "The total compliance of the EED has saved billions of euros in energy savings across Europe through companies with 5% to 15% reduction in energy consumption," the European Commission says.

The EISA conditions the use of energy-efficient practices and technologies by federal agencies. More so, EISA promotes the growth of energy-efficient appliance and vehicle standards. EISA, as reported by the U.S. Department of Energy, has already inspired innovation in the realm of energy-efficient technologies, saving the federal supply chain from 10% of its energy used.

## Availability of Financial Incentives and Subsidies

Governments often provide financial incentives and subsidies to encourage companies to invest in green technologies and practices. Such incentives include tax credits, grants, and rebates.

For instance, the federal government of the United States has had the capability to offer attractive tax breaks for investments in renewable energy systems, such as wind turbines and solar panels. Bloomberg NEF 2023 indicates that ITCs have been essential in making supply chains more compelling by stepping up their adoption of renewable energy technologies, resulting in a marked growth in clean energy capacity and a reduction in energy costs [32].

Specifically, the Renewable Energy Incentive Program of Australia provides financial incentives to enterprises for implementing renewable energy and energy-saving technology. Notably, the Australian Renewable Energy Agency has emphasized that such a program has contributed to an increase in the adoption of renewable energy among businesses participating in the program, with some experiencing a boost in uptake as high as 12%, thereby saving enormous amounts of energy while generating fewer emissions [33].

## Facilitating Transparency and Reporting

The laws that enforce transparency and reporting requirements can also influence the efficiency of the supply chain. Incentives for disclosing energy usage, carbon emissions, and sustainability practices encourage companies to adopt better practices and make enhancements in their energy performance [34].

Another significant voluntary reporting framework is the Carbon Disclosure Project, which requires companies to publicly disclose their carbon emissions and energy usage [35]. Such companies have, over the last three years, managed to increase their energy efficiency by an average of 10% due to increased transparency and accountability, according to CDP reports.

On the contrary, the Global Reporting Initiative provides a more detailed framework for companies and businesses to report their environmental performance, including energy usage, efficiency, and management. According to a 2021 survey conducted by the Sustainability Reporting Network, the adoption of GRI standards among best practices in energy management within companies increased by 15%. A basis for increased transparency has been documented, with a focus on reporting.

## Regulatory Pressure as an Innovation Driver

There are regulatory pressures, which may also spur innovations in efficiency; for instance, requiring companies to come up with new technologies and ways of doing things. For example, government regulations can push a country into ambitious energy efficiency targets or reductions in carbon emissions and catalyze investments in R&D.

Some notable examples include the Clean Power Plan, established during the Obama administration and later replaced by the Affordable Clean Energy Rule [36]. The rule aimed to lower greenhouse gas emissions from power plants. Despite an implementation challenge, the policy strongly encouraged a surge in innovation related to clean energy technologies and energy efficiency in the power sector. Brookings Institution (2020) reported that investment in clean energy R&D went up by 20% owing to the Clean Power Plan.

A broader scale finds the Paris Agreement, with its plans for reduction in greenhouse-gas emissions worldwide, pushing almost every company on the planet toward innovation and the development of energy-efficient technologies. In fact, according to the International Energy Agency (2022), the Paris Agreement has spurred the development and deployment of energy-efficient technologies, in particular, in the transport and manufacturing sectors.

Policy and regulation are key factors that encourage supply chain efficiency through the implementation of mandatory standards, financial incentives, transparency, and innovation. They drive companies to improve their energy management practices in pursuit of further sustainability goals.

## CASE STUDY

### Opportunities, Challenges, and Strategic Recommendations for Boosting Energy Efficiency in Indian Pharmaceutical SMEs

A critical segment of India's industrial economy and a key provider of medical equipment is the Indian pharmaceutical industry. Yet, sometimes the manufacturing process of pharmaceuticals is energy-intensive, especially in terms of chemical synthesis, clean room operation, and HVAC (Heating, Ventilation, and Air Conditioning) systems. Enhancing energy efficiency in Small and Medium Enterprises (SMEs) within this industry is crucial in light of the rising energy costs and the urgency of the climate crisis. Increasing awareness of energy efficiency has yet to be matched with effort, and implementation remains patchy in the face of a multifaceted array of barriers.

A brief case study of Mumbai-based MediCure Pharmaceuticals Pvt. Ltd. (Speculative)

The 150-employee Mumbai SME MediCure Pharmaceuticals produces generic drugs for local sale. The company's electricity bills were increasing, and occasional audits revealed inefficiencies in the production facility and heating, ventilation, and air conditioning units. Initially, MediCure was reluctant to invest in energy-efficient retrofits because of limited internal technical resources and limited funds.

However, the company implemented Variable Frequency Drives (VFDs) on pumps and installed a Building Management System (BMS) after receiving a grant from the Bureau of Energy Efficiency (BEE) and technical assistance from an adjacent Energy Service Company (ESCO). MediCure achieved a 14% boost in system reliability and a 22% decrease in energy cost within one year [37].

### Methods of Measurement (Energy KPIs)

MediCure measured the following Energy Key Performance Indicators (KPIs) for energy efficiency improvements [38]:

- SEC, or specific energy consumption

Energy Cost per Unit of Revenue: This metric measures the cost of energy as a percentage of revenue. The energy consumed per unit product (kWh/unit product) decreased from 1.65 to 1.28 following the intervention.

- Carbon Intensity → kg $CO_2$ produced per production unit — useful for sustainability and ESG reporting
- Clean Energy Dashboards → Facilitated ongoing monitoring and data-driven optimization

### SME Adoption Recommendations

- Finance and Incentives Accessibility → Streamline government programs (*e.g.*, BEE, SIDBI energy efficiency initiatives)
  - Promote low-interest financing and capital rebates
  - Set up regional energy training centers for SMEs
  - Organize technical workshops specific to industries
  - Partner with ESCOs
  - Encourage risk-sharing through performance contracting
  - Utilize CSR frameworks
  - Large pharmaceutical corporations can mentor or co-invest in energy efficiency upgrades for their SME suppliers

## CONCLUSION

A partnership that utilizes technological enablement and data orientation in its practice now becomes the only means of achieving significant energy savings through their supply chains, both for businesses and policymakers. For businesses, this would mean putting energy efficiency at the heart of their operations, as well as close engagement with all stakeholders and partners. For policymakers, this would amount to supportive regulation by appropriate systems that promote cooperation, innovation, and long-term sustainability. Together, these activities will contribute to building more resilient and energy-efficient supply chains in support of the global sustainable development goals.

# REFERENCES


[1] R. Baldwin, and A. Venables, *NBER working paper series spiders and snakes: offshoring and agglomeration in the global economy,* 2010. [Available from: http://www.nber.org/papers/w16611]

[2] R. Malhotra, and G. Taneja, "Comparative analysis of two stochastic models with varying demand", *Int. J. Appl. Eng. Res.,* vol. 10, no. 17, pp. 37453-37460, 2015.

[3] A. Anca, *Logistics and Supply Chain Management: An Overview,* 2019. [http://dx.doi.org/10.2478/sbe-2019-0035]

[4] C.P. Bown, "The US–China trade war and Phase One agreement," J. Policy Model., vol. 43, no. 4, pp. 805–843", *Online (Bergh.),* 2021. [http://dx.doi.org/10.1016/j.jpolmod.2021.02.009]

[5] bp, "Statistical Review of World Energy 2022," 2022. [Available from: https://www.bp.com/content/dam/bp/business-sites/en/global/corporate/pdfs/energy-economics/statistical-review/bp-stats-review-2022-full-report.pdf]

[6] Chopra, S., & Meindl, P. Supply Chain Management: Strategy, Planning, and Operation (6th ed.). Pearson, 2016. [http://dx.doi.org/10.1007/978-3-8349-9320-5_22]

[7] Energy star, *Energy star for the residential sector.* n.d. [Available from: https://www.energystar.gov/sites/default/files/asset/document/2020_About_HOMES_032520_508%20v3.pdf]

[8] A. Cherrafi, S. Elfezazi, A. Chiarini, A. Mokhlis, and K. Benhida, "The integration of lean manufacturing, six sigma and sustainability: A literature review and future research directions for developing a specific model," J. Cleaner Prod., vol. 139, pp. 828–846", *Online (Bergh.),* 2016. [http://dx.doi.org/10.1016/j.jclepro.2016.08.101]

[9] H. Kaur, R. Malhotra, "Profit Analysis of a System of Non-Identical Units with Varying Demand". Palestine Journal of Mathematics. vol 14 (Special Issue I) 2025, pp. 183–197.

[10] J.T. Mentzer, W. DeWitt, J.S. Keebler, S. Min, N.W. Nix, C.D. Smith, and Z.G. Zacharia, "Defining supply chain management," J. Bus. Logist., vol. 1, pp. 1–25", *Online (Bergh.),* 2011. [http://dx.doi.org/10.1002/j.2158-1592.2001.tb00001.x]

[11] U. Jüttner, H. Peck, and M. Christopher, "Supply chain risk management: Outlining an agenda for future research," Int. J. Logistics: Res. Appl., vol. 6, pp. 197–210", *Online (Bergh.),* 2003. [http://dx.doi.org/10.1080/13675560310001627016]

[12] Unilever Climate Transition Action Plan updated 2024. [Available from: https://www.unilever.com/files/8b5df5f6-cb90-40fd-9-91-38d06905d81d/unilever-climate-transition-action-plan-updated-2024.pdf]

[13] E. Cagno, and A. Trianni, "Evaluating the barriers to specific industrial energy efficiency measures: An exploratory study in small and medium-sized enterprises," J. Cleaner Prod., vol. 82, pp. 70–83", *Online (Bergh.),* 2014. [http://dx.doi.org/10.1016/j.jclepro.2014.06.057]

[14] P. Rohdin, P. Thollander, and P. Solding, *Barriers to and drivers for energy efficiency in the Swedish foundry industry*, 2007. [http://dx.doi.org/10.1016/j.enpol.2006.01.010]

[15] P. Thollander, J. Palm, and P. Rohdin, *"Categorizing barriers to energy efficiency – an interdisciplinary perspective," in Energy Efficiency.* Sciyo, 2010. [Online] [http://dx.doi.org/10.5772/9828]

[16] J. Palm, and F. Backman, *Energy efficiency in SMEs: overcoming the communication barrier*, 2020. [http://dx.doi.org/10.1007/s12053-020-09839-7]

[17] P. Thollander, J. Palm, and P. Rohdin, *Categorizing barriers to energy efficiency: An interdisciplinary perspective.*www.intechopen.com
[http://dx.doi.org/10.5772/9828]

[18] D. Ivanov, and A. Dolgui, "Viability of intertwined supply networks: Extending the supply chain resilience angles towards survivability. A position paper motivated by COVID-19 outbreak," Int. J. Prod. Res., vol. 58, pp. 2904–2915", *Online (Bergh.),* 2020.
[http://dx.doi.org/10.1080/00207543.2020.1750727]

[19] S. Seuring, and M. Müller, "From a literature review to a conceptual framework for sustainable supply chain management," J. Cleaner Prod., vol. 16, no. 15, pp. 1699–1710", *Online (Bergh.),* 2008.
[http://dx.doi.org/10.1016/j.jclepro.2008.04.020]

[20] A. Wieland, R.B. Handfield, and C.F. Durach, "Mapping the landscape of future research themes in supply chain management," J. Bus. Logistics, vol. 37, no. 3, pp. 205–212", *Online (Bergh.),* 2016.
[http://dx.doi.org/10.1111/jbl.12131]

[21] R. Locke, F. Qin, and A. Brause, "“Does monitoring improve labor standards? Lessons from Nike,” MIT Sloan School of Management", [Available from: http://ssrn.com/abstract=916771]

[22] M.E. Porter, and C. Linde, "Toward a new conception of the environment-competitiveness relationship", *J. Econ. Perspect.,* vol. 9, no. 4, pp. 97-118, 1995.
[http://dx.doi.org/10.1257/jep.9.4.97]

[23] R.S. Pindyck, "The long-run evolution of energy prices", *Energy J. (Camb. Mass.),* vol. 20, no. 2, pp. 1-27, 1999.
[http://dx.doi.org/10.5547/ISSN0195-6574-EJ-Vol20-No2-1]

[24] H.L. Lee, "Aligning supply chain strategies with product uncertainties," California Manage. Rev., vol. 44, no. 3, pp. 105–119", *Online (Bergh.),* 2002.
[http://dx.doi.org/10.2307/41166135]

[25] H. R. Varian, Big data: New tricks for econometrics," *J. Econ. Perspect.,* vol. 28, no. 2, pp. 3–28, 2014.
[http://dx.doi.org/10.1257/jep.28.2.3]

[26] S. Dhawan, Online learning: A panacea in the time of COVID-19 crisis," *J. Educ. Technol. Syst.,* vol. 49, no. 1, pp. 5–22, 2020. [Online]. Available 10.1177/0047239520934018

[27] T. Mikropoulos, and A. Natsis, *Educational virtual environments: A ten-year review of empirical research (1999-2009).* vol. Vol. 56. Computers & Educ, 2011, pp. 769-780. [Online]
[http://dx.doi.org/10.1016/j.compedu.2010.10.020]

[28] M. Geissdoerfer, P. Savaget, N.M.P. Bocken, and E.J. Hultink, "The circular economy – A new sustainability paradigm?" J. Cleaner Prod., vol. 143, pp. 757–768", *Online (Bergh.),* 2017.
[http://dx.doi.org/10.1016/j.jclepro.2016.12.048]

[29] S. K. Srivastava, Green supply-chain management: A state-of-the-art literature review," Int. J. Manage. Rev., vol. 9, no. 1, pp. 53–80, 2007.
[http://dx.doi.org/10.1111/j.1468-2370.2007.00202.x]

[30] International Energy Agency, *Energy Effic.,* 2020.

[31] International Renewable Energy Agency, *Global Energy Transformation: A Roadmap to 2050,* 2018. [Available from: www.irena.org]

[32] International Renewable Energy Agency, *World energy transitions outlook 1.5°C pathway executive summary.*

[33] *Numerical analysis of fatigue failure.* vol. 2357. , 2022, p. 100009.R. Malhotra, A. Soni, and A. Goyal, AIP Conf. Proc. [Online]
[http://dx.doi.org/10.1063/5.0080979]

[34] R. Malhotra, and A. Jain, *"IoT-based improved mechanical design for elevator cart," in Applied Intelligence in Human-Computer Interaction.,* S. Bansal, P.C. Sharma, A. Sharma, J-R. Chang, Eds., 1st ed. CRC Press, 2023. [Online]
[http://dx.doi.org/10.1201/9781003415466]

[35] R. Malhotra, *Reliability and availability analysis of a standby system with activation time and varied demand.* Eng. Reliab. Risk Assess, 2022, pp. 35-40. [Online]
[http://dx.doi.org/10.1016/B978-0-323-91943-2.00004-6]

[36] *Design and construction of solar shed.* vol. 2220. , 2020, p. 140061.R. Malhotra, and B. Singh, AIP Conf. Proc. [Online]
[http://dx.doi.org/10.1063/5.0001157]

[37] P. Thollander, and J. Palm, *Improving Energy Efficiency in Industrial Energy Systems: An Interdisciplinary Perspective on Barriers, Energy Audits, Energy Management, Policies, and Programs.* Springer-Verlag London, 2013.
[http://dx.doi.org/10.1007/978-1-4471-4162-4]

[38] Bureau of Energy Efficiency (BEE), [Available from: https://beeindia.gov.in]

# CHAPTER 8

# Case Studies and Examples

**Shilpy Bhullar**[1,*]

[1] *Schepens Eye Research Institute, Massachusetts Eye and Ear Infirmary of Harvard Medical School 20 Staniford St, Boston 02114, MA, USA*

**Abstract:** The chapter "Case Studies and Examples" provides a compelling exploration of successful sustainable energy projects, offering in-depth analyses of pivotal case studies and exemplary initiatives. Through meticulous examination, this chapter illuminates the intricate tapestry of sustainable energy deployment, shedding light on the technologies, challenges, and triumphs that characterize these transformative endeavors. Two types of case studies have been discussed - "Type 1: Renewable Energy Integration in Urban Settings" and "Type 2: Community-Based Microgrid Development". Type 1 study thoroughly discusses the installation of energy-saving solutions such as solar photovoltaic systems and wind turbines to meet the growing energy demands of buzzing megacities. It explains the importance of implementing regulatory frameworks, considering urban planning solutions, and engaging stakeholders in decision-making, offering valuable insights into the successful amalgamation of sustainable energy technologies within majorly populated cities. Type 2 study deals with brainstorming the intricacies, executing innovative technology into functioning, and strategizing ways to maximize community engagement that lay the foundation for successful development of the microgrid. It details the hardships faced, including technical difficulties and monetary constraints, and enlists the successful steps undertaken that led to the fruition of different projects. These case studies exemplify the impact of sustainable energy solutions that are playing a key role in meeting the upcoming energy demands. This chapter provides the readers with a deeper understanding of the challenges and triumphs associated with realizing sustainable energy projects, showcasing to them the importance of informed decision-making and innovation.

## INTRODUCTION

Our planet is facing challenges such as climate change, global warming, pollution, and the depletion of fossil fuels, due to which the requirement for sustainable energy solutions has become more prominent today thanever before. According to the Intergovernmental Panel on Climate Change (IPCC), unless greenhouse gas missions are substantially minimized, the global temperatures would rise by

* **Corresponding author Shilpy Bhullar**: Schepens Eye Research Institute, Massachusetts Eye and Ear Infirmary of Harvard Medical School 20 Staniford St, Boston 02114, MA, USA; E-mail: sbhullar@meei.harvard.edu

**Reetu Malhotra, Pooja Mahajan, Manish Singla & Ramesh Kumar (Eds.)**

more than 1.5 degrees Celsius above pre-industrial levels by the year 2030, which would have severe consequences [1]. This will affect the urban areas where more than half of the world's population resides and consumes approximately 70% of the global energy.

Renewable energy sources, including solar energy, wind energy, hydroelectric energy, geothermal energy, and biomass, present a solution to eradicate the growing menace of climate change and depletion of fossil fuels. These technologies, if incorporated into the energy grid, not only ensure energy security but also facilitate a transition toward a low-carbon footprint by reducing greenhouse gas emissions. Recent developments in Photovoltaic (PV) efficiency, wind turbine design, and energy storage systems strengthen the potential for these renewable sources of energy to sustainably tackle current energy demands. Furthermore, interdisciplinary research into smart grid technologies and decentralized energy systems demonstrates the ease of optimizing energy distribution and its consumption, thereby addressing the challenges of sporadicness and lack of flexibility, which are inherent to renewable sources. With the growing energy demand among the masses and ambitious targets set forth by the international climate agreements, a continuous investment in research and development is crucial to unlock the full potential of renewable energy sources to ensure a resilient and sustainable future.

Renewable energy sources offer huge economic benefits not only in terms of financial metrics but also in terms of profound societal influence. Transitioning to renewable sources of energy creates more jobs, ranging from manufacturing of the equipment to their installation and further research, thereby helping local communities to grow and reducing unemployment. Such jobs will create a more skilled workforce as more training programs will be launched to enhance the skills of the workforce to match the growing demands. Furthermore, this will reduce the reliance on fossil fuels, ensuring energy independence, improving trade balance, and also safeguarding the economy from vulnerable and sporadic global energy prices. Deploying renewable energy technologies reduces the carbon footprint, and the operational costs involved also reduce with time. Once the renewable energy sources become fully functional, the cost of energy generation reduces considerably over time because the stocks of solar energy, wind energy, *etc.*, are immense. With research and development, we can further refine our technology to reap maximum benefits. This results in significant savings both for the consumers and the government or businesses, enabling the stakeholders to budget expenses in other areas of the economy. One cannot ignore the major environmental benefits that can be harvested, such as improved air quality and reduced healthcare costs. Thus, renewable technology not only fosters sustainable economic growth but also an improved quality of life.

The core of any developing or developed community lies in its urban areas. This is the reason that incorporating renewable energy sources and community microgrids within urban settings has the potential to reap maximum benefits. This involves transitioning from centralized fossil-fuel-dependent systems to decentralized sustainable systems to harness energy [2]. Community-based microgrids empower local communities to manage their energy requirements sustainably with reduced reliance on conventional grids, which are often susceptible to price rises and outages [3].

This renewable energy-based sustainable approach has multiple benefits - (**i**) there is a significant reduction in carbon emission and environmental degradation, (**ii**) energy costs are greatly reduced, (**iii**) employment opportunities arise for the local population, (**iv**) energy independence is ensured, (**v**) social security and satisfaction is ensured, (**vi**) affordable energy is available for underserved communities, and (**vii**) there is always a scope for improvement in the technology being used with extensive research and development. Community-based microgrids offer a more localized nature of the microgrids, which can be consistently improved for efficiency and reliability. Overall, the introduction of both renewable sources of energy and community-based microgrids in urban environments presents not just an option but an imperative solution to tackle the persistent threats posed by global warming and drastic climate change and to ensure a safe, secure, and sustainable future for the upcoming generations.

This chapter delves deeper into studying the intricacies involved in the successful installation and functioning of sustainable energy-based projects through a number of compelling case studies and initiatives. The chapter discusses the technology employed, obstacles faced, and factors responsible for the success of the projects discussed so far. The focus of this chapter lies in two significant areas: **(i)** integration of renewable sources of energy and **(ii)** community-based microgrids in urban settings. The case studies discussed hold special importance because the projects represent examples that align harmoniously well with the United Nations Sustainable Development Goals (UN SDGs), as shown in Fig. (**1**).

**Fig. (1).** Sustainable Development Goals (SDGs) of the United Nations.

## CASE STUDY TYPE 1: RENEWABLE ENERGY INTEGRATION IN URBAN SETTINGS

This section focuses on the introduction of renewable energy sources to urban settings. It unravels the challenges faced and the triumphs by outlining the factors that were found to be responsible for the successful implementation of the projects. It unfolds the complexities of regulatory frameworks, community involvement, government policies, technological intervention, and careful planning in order to offer successful integration of renewable sources of energy within densely populated areas.

I. **City of Soria, Spain:** This example mentions the successful implementation of the sustainable energy project in the city of Soria in Spain. The project was based on installing solar energy panels in municipal buildings. Three options were analyzed: (**i**) PV-based solar energy system on rooftops, (**ii**) hybrid (thermal and PV-based) solar energy, and (**iii**) a PV generation system with hydraulic storage. Key challenges noted during the implementation of this project include adapting the technological innovations to the existing infrastructure, ensuring economic feasibility, and meeting the energy demand. The project was able to meet the electrical and thermal demands in perfect alignment with the sustainable and integrated urban development strategies. The highlighting feature was that the project promoted self-consumption owing to the contributing factors such as advanced PV technology, high solar potential, and supportive government reforms. In Sustainable Energy Planning (SEP), approaches like analyzing patterns of energy consumption, optimizing PV systems installation, and using software such as PVSyst and HOMER PRO were used to ensure an effective integration and its management. Integration involves planning for how the new technology will be implemented in the already existing settings, such as analyzing the orientation of the buildings, rooftops, the already existing electrical wiring and setups, and how the buildings are designed, *i.e.*, whether neighbouring buildings shadow each other or interpreting different times when each of the buildings is fully exposed to the sun. Management involves monitoring energy production and ensuring that the production is capable of meeting the energy needs, economic feasibility, and regulatory compliance. This project has exemplified how a successful sustainable energy initiative can be achieved with planning strategically, technological integration, and supportive government policies [4].

II. **Nigeria:** This is an example of a longitudinal study conducted on infrastructural development in Nigeria to interpret the relationship between investment and economic development between the time period from 1996 to 2019. This study concluded that both economic and social infrastructure are responsible for reducing poverty and the overall upliftment of society. A strategically planned infrastructural development and significant improvement in the economic outcomes and living standards of the society were observed. The impact of a well-planned infrastructural development on poverty alleviation was confirmed through the Autoregressive Distributed Lag (ARDL) bounds testing approach and the Granger causality test. There was a positive correlation between increased infrastructure spending and reduced poverty. Infrastructural development included both economic infrastructure, such as transport, communication, *etc.*, and social infrastructure, such as health and education. Thus, the results concluded a bidirectional relationship between infrastructure and poverty, *i.e.*, improving infrastructure can reduce poverty, and reducing poverty can enhance infrastructural development. The study concluded that Nigeria must continue focusing on strategic infrastructural investments in order to achieve better human development outcomes, as it is a factor responsible for reducing poverty and improving the lifestyle of the inhabitants [5].

III. **Reunion Island:** The study was conducted in Reunion Island to evaluate scenarios for plastic waste management based on the indicators from the UN SDGs. The Business as Usual (BAU) criteria scored poorly across all the indicators, thereby mandating urgent steps to be taken to rectify it. The Zero Waste Plan (ZWP) and the European Circular Economy Package (ECEP) strategies were implemented and showed better results, with some limitations, as local challenges were the most difficult to tackle. A new scenario, New Technology for Plastic Recovery (NTPR), proved to be the most successful, proposing that strong local policies are imperative to ensure effective plastic waste management. The study also focused on ensuring resource management. Thus, the NTPR policy adopted seemed to offer promising solutions for sustainable plastic waste management as it addressed local needs while meeting both environmental and economic goals [6].

IV. **Cities of Rio de Janeiro, Jakarta, Vancouver, New York City, Utrecht, Cape Town, and Stirling:** This case study mentions successful sustainable energy projects in the cities of Rio de Janeiro, Jakarta, Vancouver, New York City, Cape Town, Utrecht, and Stirling, advocating the importance of city-driven initiatives and local community engagement. Different technologies that were offered as part of energy sources include electric buses, solar PV, solar power technology with battery storage, Vehicle-t--Grid (V2G) systems, Electric Vehicle (EV) charging, and renewable power purchase agreements. The initial obstacles were regulatory barriers, financial complexities, grid congestion, and social involvement. But, supportive government policies, strong community involvement, innovative financial tactics, and international collaboration led to the triumph of these proposals. Thus, strategic and integrated planning, data-driven decision making, and inclusive approaches for maximum involvement proved to be the prime factors responsible for achieving sustainable energy projects in urban settings [7].

V. **Okanagan, British Columbia, Canada:** This study focuses on a project proposed in a residential neighborhood in Okanagan, British Columbia, Canada. The key technological components involved were solar PV systems, geothermal heating, and a waste-to-energy incineration plant. However, high investment costs and requirements for technical expertise were observed as major challenges in planning a renewable energy setup in this area. The factors contributing to the success of the project include government policies and incentives, strategic investment for cost recovery, and stakeholder involvement. The operating team in this project prioritized mitigating energy consumption through systematic amalgamation of technology and methodology and assimilating locally available renewable sources of energy to meet growing energy demands. Environmental and economic impacts of the energy sources were carefully researched to choose the optimal renewable source, which can offer maximum benefits. Government, local community, and developers were engaged to ensure mass acceptance and overall success of the projects. The study underscores various approaches in SEP, such as scenario-based analysis, emphasizing energy conservation, stakeholders' involvement, and putting local resources to the best possible use [8]. The proposed factors led to the success of the project despite the initial challenges.

VI. **City/area not specified:** This example does not mention any specific area but discusses the concept of successful integration of sustainable energy projects within urban settings by focusing on building clusters of renewable energy sources. The key city/area discussed was not specified, but the research emphasized urban settings in general. Technologies discussed include Building-Integrated Photovoltaics (BIPV), Solar Thermal Façades (STF), and heat pumps. The challenges faced consist of high processing costs of existing models, uncertainties in computational methodologies, and barriers in energy system integration. But, accurate modelling, multi-criteria analyses (3E and 4E), and adaptive building envelope solutions contributed towards the success of the proposals suggested. Various approaches used in SEP include top-down and bottom-up methods, micro-simulation, and Urban Building Energy Simulations (UBES). Integration and management strategies focused on energy efficiency retrofits, local renewable energy supply, and Demand Response (DR) programs. In an urban setting, the successful implementation of sustainable energy projects hinges on the integration of renewable energy sources within building clusters. Technologies such as BIPV, STF, and heat pumps are pivotal. However, challenges like high processing costs, uncertainties in computational methodologies, and barriers in energy system integration emerge and must be addressed. An effective and efficient SEP employs both top-down and bottom-up methods. Overall, the effective integration and management strategies resulted in ensuring a resilient and efficient urban energy system [9].

VII. **China's Smart City Project:** The case study examined the successful implementation of sustainable energy projects in China's Smart City Pilot Program, focusing on coastal, large, and resource-based cities. These projects utilized advanced digital technologies such as smart grids, intelligent transportation systems, real-time tracking, smart scheduling, contactless payment systems, and eco-friendly modes like electric buses. The project faced challenges such as high initial costs, infrastructure costs, privacy and security concerns, volatility, and concern for substantial investments; however, factors such as financial inclusion, effective resource management, digital technology, and green innovation have managed to sustain the project. The unification of innovative technologies and effective approaches has promoted the sustainable development of renewable energy projects in metropolitan areas by effectively addressing environmental concerns and meeting energy demands [10].

VIII. **Emilia-Romagna region of Italy:** This example mentions the successful installation and performance of sustainable energy projects through three case studies performed in the Emilia-Romagna region of Italy - Bologna, Modena, and Ravenna. These cities incorporated Sustainable Energy and Climate Action Plans (SECAPs) and General Urban Plans (GUPs) to improve energy competence and buoyancy. The technologies incorporated PV panels, wind turbines, and energy-efficient buildings. The challenges faced included incorporating new technologies into existing infrastructures, addressing poverty, involving the public, and promoting awareness. The factors that contributed toward the success of the projects were using government incentives for energy generation, renovation, and preservation, meeting high-energy requirements for redevelopment, and promoting energy communities. The integration of energy resilience strategies into urban planning is crucial, focusing on land use, urban morphology, governance, individual behavior, and socio-demographic aspects. Approaches used in SEP include mapping energy consumption, promoting zero-energy districts, and creating local distribution networks. The comparative analysis of SECAPs and GUPs highlights the importance of strategic planning, stakeholder engagement, and continuous evaluation to achieve sustainable urban energy systems [11].

IX. **Barcelona and Singapore's Marina Bay Smart District:** This case study mentions successful smart city initiatives in Barcelona and Singapore's Marina Bay Smart District. These projects utilized solar panels integrated into public spaces, IoT infrastructure, real-time data analytics, and smart grids. The primary challenges faced include limited space, shading issues, aesthetic considerations, and high energy demand. But the measures that turned out to be the success factors include design innovations, efficiency enhancements, community engagement, and international collaboration [12].

X. **Diu, India, and Georgetown, Texas, USA:** This case study highlighted Diu, India, and Georgetown, Texas, USA as examples of successful sustainable energy projects. Diu, India, utilized solar energy from two solar parks and rooftop panels on public institutions. Georgetown, Texas, enrolled a combination of wind and solar energy for the project. Diu faced issues that required support from the central government for land acquisition, planning, financing, and implementation. The success in Diu was established through government support. Georgetown faced challenges with uncompetitive prices and excessive and unregulated energy production, which led to financial losses. The success of the energy project launched in this US town thrived on long-term fixed-cost contracts and zero carbon risk. The common success indicator in both projects involved SEP, integration, and management.

Planning - This involves setting clear renewable energy goals and getting the necessary support and finances to turn those anticipated goals into reality.

Integration - This involves including new and innovative renewable energy technology into already existing infrastructure, ensuring its smooth transition into the city settings, and making the best possible use of the available resources.

Management - This involves optimizing and maintaining energy systems, ensuring economic feasibility, and addressing concerns like excess energy production, energy loss and wastage, and getting better incentives for the services offered. Both studies utilized approaches based on effective planning, integration, and management [13].

XI. **Dezhou, China:** This case study explains the success behind the sustainable energy project setup in Dezhou, China. The Dezhou Economic Development Zone was developed in the city to establish a solar energy industry, for which technologies such as PV systems, solar thermal energy systems, and solar water heaters were introduced. This project also faced challenges such as a poorly developed financial sector, the small size of the solar energy industry, and an immense shortage of skilled workforce. The major factors that proved to be responsible for the success of the solar energy project were investment in Research and Development (R&D), education, and technological innovation. Supportive government policies were made by the local government on financing, tax returns, land usage, and attracting investments and stakeholders toward finding innovative approaches to implement solar energy systems into existing as well as newly-built infrastructure. Various approaches used in SEP included creating clusters for private actors, public stakeholders, and research institutions, and promoting innovation and economic development through synergies. The project also emphasized the importance of local government support in promoting renewable energy industries and catalyzing economic development [14].

XII. **Seattle, USA, and Manama, Bahrain:** The study reviewed successful sustainable energy projects, focusing on the Bullitt Center in Seattle, USA, and the Bahrain World Trade Center in Manama, Bahrain. Bullitt Center - This building integrated solar panels, rainwater collection, composting toilets, and ground-source heat pumps. Navigating regulatory frameworks, ensuring system compatibility, and managing high initial costs and ongoing maintenance were the starting obstacles, but detailed planning, collaboration with local authorities, and a holistic approach to sustainability led to the overall success of the project. Bahrain World Trade Center - This project used wind turbines integrated into the building design. Optimizing turbine efficiency in urban environments, maintaining turbines at height, and managing noise and vibration were the initial challenges, but success was achieved through innovative design, strategic location, and advanced maintenance plans. The approaches utilized by both projects towards achieving SEP include Integration - Combining multiple renewable technologies like solar, wind, and geothermal, Management - Ensuring system compatibility, regular maintenance, and compliance with regulations, and Planning - Detailed design considering local climate, resources, and collaboration with authorities [15].

XIII. **Italy:** The document discussed three case studies of Renewable Energy Communities in Italy, focusing on their development, challenges, and success factors.

**Case Study: Magliano Alpi (Piedmont, Northern Italy)**

PV systems, smart meters, and EV charging stations were the technologies used. Key challenges in this project were building trust among citizens, integrating all the proposed technologies, and managing energy production and consumption. Strong collaboration between the Municipality, Polytechnic of Turin, and local technicians, use of a management platform (Energy4Com), and active citizen participation led to the success of this project.

**Case Study: Ferla (Sicily, Southern Italy)**

PV systems were deployed. The project faced legal and regulatory hurdles initially, but strong community feeling, involvement of local authorities and non-government organizations, and clear communication strategies proved to benefit this project.

**Case Study: East Naples (Campania, Southern Italy)**

This project also deployed PV systems. Addressing energy poverty, social cohesion, and bureaucratic obstacles posed an initial threat to the project.

However, support from Legambiente Campania, Fondazione Famiglia di Maria, and local women who focused on social justice and energy equity supported the progress of this proposal.

The approaches which involved SEP, integration, and management in all the above-mentioned projects include Actor-Network Theory (ANT), which was used to analyze the alliances between social actors and technologies, macro and micro levels, which combined quantitative and qualitative methods to monitor and analyze RECs, and community involvement, which emphasized the role of local authorities, non-government organizations, and citizens in the planning and management process [16].

XIV. **Kuortane, Finland:** The article discussed a successful sustainable energy project conducted in the Olympic Training Center in Kuortane, Finland. The project aimed to replace fossil energy with renewable energy sources to achieve zero carbon emissions. The technologies used included geothermal energy, biomass energy, and solar energy. The project faced challenges such as the high upfront costs of geothermal energy, the intermittency of renewable energy sources, and the need for a reliable energy supply to meet the diverse demands of the training center's buildings. The factors that proved to be indispensable behind the success of the project involved integrating solar and geothermal sources, using a generic and integrated framework for Multi-Energy Hybrid Systems (MEHS) configurations, and receiving support from local authorities and private foundations. The importance of economic analysis, site-specific feasibility assessments, and modeling uncertainty considerations was highlighted through the study. The study emphasized using MEHS to ensure the reliability of renewable energy in alignment with planning, integration, and management. The effective strategies involved predicting energy demand models, analyzing the market's energy consumption and requirement, and splitting space heating and hot water usage. Different methodologies used included economic analysis, stochastic analysis, feasibility analysis, and optimizing proposed models. The emphasis was laid on optimally using locally available energy sources, incorporating space-time features, and community involvement and effects for successful arrangement. The project focused on the fact that detailed decision-making, strategic framework, and advanced modeling approaches are crucial for the effective functioning of sustainable energy projects in urban settings [17].

XV. **European cities:** This case study summarizes the success story of sustainable energy projects set up in 362 notable European cities spanning across Turkey, France, Belgium, Sweden, Greece, Norway, Portugal,

Norway, and Spain. The goal behind the projects was to extend the usage of renewable energy to tackle alarming environmental conditions. The renewable energy sources, as discussed above, were employed in this project as well, such as wind turbines, solar PV, solar thermal systems, geothermal energy, and biomass. Lack of government support, scarce financial funds, regulatory barriers, and implementing new renewable technology into an existing infrastructure proved to be major challenges. But, promoting public-private partnerships, community involvement, economic incentives, and friendly policies resulted in the overall success of the projects, and all these factors seemed to effectively contribute towards the trio of planning-integration-methodology approaches, which are supposed to influence the successful implementation of sustainable energy projects within urban settings [18].

## CASE STUDY TYPE 2: COMMUNITY-BASED MICROGRID DEVELOPMENT

This section of case studies also discusses different projects and their success stories, similar to the previous section, but here the focus is on a community-driven, community-based, resilient, and self-sustaining microgrid system. As the title mentions 'Community-based Microgrid', the success of all the projects discussed in this section relies strongly on community support. These are the projects that were proposed for the community and were realized mainly because of the community's involvement, in addition to other intricacies such as planning and innovation. Each case study mentions the complexities faced and the solutions that finally propelled the projects to fruition.

I. **Dubai, UAE, and Hemberg, Switzerland:** Case studies from Dubai, UAE, and Hemberg, Switzerland - two drastically different geographical locations- were analyzed. Various urban configurations were represented using urban archetypes to investigate the urban energy demand and the proposed energy systems design. The proposal employed an energy hub comprising MCES with renewable energy sources such as solar PV panels and wind turbines, with energy storage and energy conversion technologies. The major issues faced were managing complex urban configurations, implementing renewable energy systems into microgrid networks, and tackling variations in energy demands. However, optimal urban planning, optimization of energy systems to meet varying energy demands, reducing energy consumption, and ensuring cost-effectiveness resulted in the success of the proposed projects [19].

II. **Bangladesh:** This study implemented renewable energy-driven community microgrids in Bangladesh. Both solar PV panels and wind turbines were installed to make optimum utilization of the available solar and wind energy. Initially, the project faced challenges such as high costs, regulatory barriers, technical complexities, and social unacceptance. But introducing economic reforms, budgeting the expenses wisely, supportive government regulations, advancing innovative technology, and maximizing community involvement led to the success of the project. The study focused on combining renewable energy sources, energy storage systems, and smart control systems into one entity so that maximum benefit can be obtained. SEP approaches followed in this study include community engagement and timely public meetings, supportive policies, and demand-side management for the effective functioning of the microgrid system [20].

III. **Dual-Layer Energy Management Model Study:** This example weighs upon the Dual-Layer Energy Management (DLEM) model for a community microgrid, which includes Distributed Energy Resources (DERs), a grid, and a community. Reducing costs and ascertaining effective energy management were found to be major contributing factors. The community-based microgrid was set up at a university campus, which also included commercial and residential dwellings together with solar PV systems, EV charging stations, and Battery Energy Storage Systems (BESS). Both grid-to-Vehicle (G2V) and V2G configurations were used by the microgrid for solar PV panels, BESS, and EV stations. The factors that were observed as challenges included managing the sporadicity of the renewable sources of energy, optimizing the load demand, and adjusting the EV load profile so that it can match the expectations of the EV owners. The progress of the project was associated with the state-of-the-art probabilistic modeling load profiles of EV, introducing V2G-DRS mode for EVs, and the dual-layer approach to ensure participation and involvement of all energy entities in the market. The DLEM model was designed to reduce the net-billing cost for the community and the net-operating cost of the DERs. The model used advanced modeling techniques to tackle uncertainties associated with load demand, EV load profile, and solar power generation. In this model, the amalgamation of DRS with V2G helped in minimizing load problems and ameliorating the overall performance of the microgrid [21].

IV. **Indonesia:** This example discusses the launch of a successful sustainable energy project involving PV-based microgrids in small island communities in the Maluku Province of Indonesia. PV and energy storage systems were deployed to harness solar energy to convert it into electricity. The initial days post-launch were not easy, as there were limited logistics, limited capital, and constant damage to microgrids due to geographical isolation. The project incorporated approaches governing SEP, integration, and management through the Enterprise Architecture (EA) approach and ArchiMate modeling approach to construct a structured guideline for better operation of PV-based microgrids. These approaches were validated by reviewing case studies and obtaining expert opinions, which can confirm the relevance and practicality of the project. The key factors determining the success were Operation and Maintenance (O&M) practices and engagement of the stakeholders [22].

V. **Southeast Asia:** This case study belongs to a sustainable energy project constructed in a rural region of Southeast Asia. Solar PV and battery systems were deployed in Malaysia's Rumah Panjang region to overcome the logistical challenges through timely monitoring and community involvement. Timodos micro-hydro plant is another example in the Philippines that gained significant success owing to community involvement and timely monitoring, despite high costs, affordable tariffs, and dependence on foreign funding. The Pa Mi and Lung Thuoc projects in Vietnam were successful because these projects prioritized community participation. All these three projects discussed emphasized the significance of local supply chains, effective financial models, community engagement, and effective government policies, which were also the key prospects in SEP [23].

VI. **Business Model Study:** Social capital and business models were employed to enhance the durability of community microgrids. Local Energy Markets (LEMs), energy storage devices, and DERs were included. The business model study focused on selecting an appropriate model that aligns well with the persisting economic conditions, cultural and social aspirations, and basic energy needs. Emphasis was laid on Multi-Criteria Decision Analysis (MCDA), which again implies that sustainable solutions are offered that are in sync with the economic feasibility, social synergy, and communal harmony in order to maintain optimum functioning of the community microgrids [24].

VII. **Local Energy Market:** This example of a case study documents a successful sustainable energy project that involved setting up multi-microgrids in LEMs. This project was part of the community energy trading market and had multiple microgrids exchanging energy with each other and the main grid. The challenges faced included creating an appropriate framework, ascertaining privacy protection, and supervising energy supply and demand. The approaches used in SEP-integration-management involved classifying microgrids as buyers or sellers based on energy generation by them and their consumption, using internal pricing models to increase profit and reduce grid dependency, and applying distributed algorithms to ensure privacy protection and market clearing. The success factors behind the project involved using a community manager as an intermediate coordinator, implementing a dynamic pricing model, and developing a competitive framework for energy trading. Significant profits were obtained with an increase in social welfare by 14.25% *via* a dynamic pricing model [25].

VIII. **Senegal:** This case study discussed the success of a proposed sustainable energy project through an isolated microgrid in the village of Keur, located in Senegal. The microgrid was part of rural Electrification in Senegal (ERSEN) and utilized a combination of solar PV panels and a diesel generator to supply electricity to the entire village. The project was successfully installed in the beginning, but as its operation progressed, it faced issues such as technical complications with inverters and batteries, ineffective management and maintenance of the microgrid, and increased vulnerability among the villagers due to the costs involved. The major elements contributing to the success included the involvement of stakeholders, monetary support from international agencies, and training the local technicians. Thus, a multidimensional approach toward sustainability was adopted, consisting of financial, social, cultural, technical, and institutional elements. It was also felt that the successful implementation of the project needs a continuous backup in the form of external support, long-term planning, and maximal integration of local resources to secure the future of sustainable energy projects in rural as well as remote settings [26].

IX. **Hot Springs, North Carolina, USA, and Panton, Vermont, USA:** A 2 MW solar and 4 MW/MWH battery storage microgrid was successfully launched in Hot Springs, North Carolina, USA, despite initial cost-related concerns and opposition faced by the local community due to safety concerns. Regional stakeholders, environmental groups, and the energy innovation task force provided immense support for its productivity. Local people were actively engaged, and their safety concerns were effectively addressed through various public meetings and training sessions. Similarly, another project got its success in Panton, Vermont, USA, which installed a 4.9 MW solar array and a 1 MW/4 MWH Tesla battery system. The project addressed aesthetic concerns with vegetative screening, engaged with town leaders and residents, and conducted a life-cycle analysis of greenhouse gas emissions. Supported by state agencies, the project also provided additional community benefits like energy-efficient upgrades to the town hall. Insights into SEP highlighted the importance of community engagement, offering multiple component benefits, preference for renewable energy sources, and regulatory support [27].

X. **Dhaka, Bangladesh:** The article discussed a successful sustainable energy project in Hazaribagh, Dhaka, Bangladesh. The project involved designing a community microgrid based on PV installations on rooftops, aiming to create a positive energy community. The technologies used include solar PV panels, wind turbines, and BESS. The challenges faced include frequent electricity outages, low clean energy generation, and the need for cost-effective solutions. Prime factors contributing to the project's success were the use of HOMER Pro software for system design and simulation, real-time data analysis, and the integration of renewable energy sources to reduce costs and emissions. The project achieved a cost of energy of $0.0357/kWh, which is 52% less than the current tariff rate, with a 70% renewable fraction. Various approaches in SEP included resource assessment, load profiling, and economic feasibility analysis. Integration and management strategies involved optimizing system components, ensuring energy storage, and maintaining grid connectivity to handle demand variations and ensure reliability. The project demonstrated the potential for urban areas to transition to sustainable energy solutions, contributing to environmental goals and community well-being [28].

XI. **New York City and Colorado:** The document discussed several successful sustainable energy projects, including the Brooklyn Microgrid project in New York City and the Alamosa Solar Generating Project in Colorado.

Brooklyn Microgrid - This community-based microgrid used solar panels, battery storage, and backup generators to provide reliable and affordable electricity. Technical complexity and regulatory barriers were the major issues faced by the project. But innovative technology, community engagement, and effective management led to its success.

Alamosa Solar Generating Project - This hybrid microgrid merged a large-scale solar power plant with battery backup and natural gas backup generators into a single unit. This technical integration led to high capital costs, which were challenging to manage initially, but strategic planning and robust infrastructure resulted in its overall success. Community-based ownership, empowering local communities to own and operate microgrids, which are advanced control systems, using artificial intelligence and machine learning for efficient energy management, integration of renewable sources, combining solar, wind, and other renewable energies, resilience and flexibility, and designing systems to withstand natural disasters and adapt to changing conditions were the major components which led to the overall success of these projects [29].

XI. **Peer-to-Peer (P2P) Energy Sharing:** Another report discussed a successful sustainable energy project within a community microgrid. This project involved P2P energy sharing among 100 households equipped with technologies like solar PV panels, small-scale wind turbines, EV battery storage, and Home Energy Management Systems (HEMS). The challenges faced include pricing mechanisms, dynamic supply-demand ratios, and physical network constraints. The key factors contributing to success were the dynamic P2P pricing mechanism, Monte-Carlo simulation for uncertainties, and DR programs. The project also emphasized fairness and substantial energy cost savings. In terms of SEP, integration, and management, the document highlighted the use of double auction models and analytical models for pricing, the importance of dynamic retail electricity prices, and the integration of EVs in energy trading. The HEMS played a crucial role in optimizing energy consumption and generation in real-time, ensuring efficient participation in the P2P market. The project followed a comprehensive approach to maintain stable voltage and avoid line congestion to ensure balanced and optimum energy management, including generation, consumption, and storage, while also reaping economic benefits for all the stakeholders involved [30].

XII. **General Study:** The study mentioned in this subheading discusses the concept of sustainable energy projects through the lens of community microgrids. These microgrids were designed as small-scale energy networks capable of operating both independently and in conjunction with the main grid. DERs such as solar panels, wind turbines, hydroelectric systems, and energy storage devices like batteries were used. Initially, the factors such as lack of trust, conflict, and low participation were presented as challenges that were effectively overcome by smart leadership and collective action taken to build trust and maximize community participation. Factors like ownership and governance structures were the key drivers for the successful operation of the project, but the approach was also driven by less common factors, such as aligning the design of the microgrids with the social and cultural context of the community. Tailoring ownership, physical design, and operational strategies played a major role in the successful implementation of the project as they truly reflected the community interests and aspirations. A conceptual-theoretical framework was followed to determine the key needs and considerations of the society for the microgrid installment. Various components of this framework included community type, community capability, social capital, and microgrid impact. Thus, it can be stated that the success of the project was strongly governed by the social factors [31].

XIII. **Alaska:** A southwestern Alaskan village was chosen to study community resilience through sustainable energy solutions. The village had a population of about 70 and was struggling with high energy costs, dependence on imported diesel, and climate change impacts on food and water security. The project installed a 100 kW wind turbine with dispatchable loads such as a CropBox© container farm, Lifewater™ sewage treatment system, water reuse system, and residential water heaters. The wind turbine established a capacity factor of 16.7% and a levelized energy cost of $1.15/kWh. These dispatchable loads were capable of absorbing up to 86% of their annual load, thereby effectively managing the irregularity of the wind resource. This project also demonstrated the approach of matching energy generation requirements with the community-specific loads to enhance sustainability and resilience. Integrating innovative technologies with the interests of remote communities laid the basis of this study [32].

XIV. **Philippines:** A remote fishing village in the Philippines was chosen for a sustainable energy project that involved installing an off-grid microgrid, which consisted of 62 solar home system units arranged in 10 clusters. Solar panels and batteries were employed, and for allowed energy usage, each unit paid a regular fee. The primary motive was to ensure correct dimensions of the panels and that batteries are capable of sustaining bad weather. P2P energy exchange promoted economic development by emphasizing flexible loads to improve performance, setting up more productive units, and diversifying the energy sources. Talking in terms of SEP, integration, and management, quantitative frameworks were documented to compare microgrid setups and sharing configurations, discuss the advantages of interconnected systems and agent-based models to mimic energy circulation, and assess the performance. The key performance indicators, such as improving energy sharing, reducing energy leakage, and grid heterogeneity, proved to be effective in ensuring optimum efficiency of the off-grid microgrids setup in the remote areas of the Philippines [33].

To summarize, the projects mentioned in Type I case studies focus on large-scale energy solutions in urban environments. Solar panels, wind turbines, and other renewable technologies are implemented to meet high energy demands. The projects discussed require regulatory frameworks, strategic urban planning, and stakeholder engagement. Major challenges involve infrastructure adaptation, economic feasibility, and governmental support. The projects mentioned in Type II case studies focus on smaller, decentralized energy networks designed for local communities. Again, the projects thrive on microgrids backed with solar panels, energy storage, and community participation. These projects prioritize local involvement, economic accessibility, and energy reliability. Their challenges include technological execution, financial constraints, and community engagement. Table **1** also highlights the major differences between the two case studies observed so far.

**Table 1. Comparison of the characteristics of Type I and Type II case studies.**

| Aspect | Case Study Type I: Urban Energy Integration | Case Study Type II: Community-Based Microgrid |
|---|---|---|
| **Scale** | Large-scale, city-wide projects | Localized, small-community networks |
| **Technology Used** | Solar panels, wind turbines, integrated urban systems | Solar PV, microgrids, energy storage |
| **Primary Challenge** | Infrastructure adaptation, regulatory compliance | Community engagement, financial feasibility |
| **Implementation Complexity** | Requires policy support, city-wide integration | Needs grassroots involvement, adaptable energy models |
| **Energy Consumption** | High-energy demand from dense populations | Self-sufficient, resilient localized networks |

From a business perspective, cost-saving strategies are essential to maximize operational efficiency and improve profitability. These strategies include optimal resource allocation, investing in energy-efficient technologies, minimizing waste, and streamlining the process. Such strategies significantly impact the quality of service, thereby presenting a delicate balance between efficiency and customer satisfaction. Cost-benefit analysis must be run to identify areas where costs can be effectively minimized without compromising quality, productivity, reliability, empathy, and responsiveness. With such analysis, organizations can prioritize those initiatives that will yield them maximum returns on their investment and also ensure greater customer satisfaction [34, 35]. At no point should an organization think about reducing its expenses at the cost of poor service, longer wait times, and reduced customer satisfaction. Thus, the focal point of the cost-saving strategies should be implementing a customer-centric approach. Simplifying problem-solving and troubleshooting for consumers greatly improves their overall experience with the organization. This further ensures increased customer loyalty and long-term profitability. Moreover, renegotiating contracts with suppliers and consolidating purchases can help secure bulk discounts as well. Additionally, if the employees are sensitized about cost-saving, then this fosters innovative thinking among them as they strive to look for cost-effective yet innovative methodologies, which further help in reducing the expenses. All these measures help businesses establish their strong foothold in the market, ensuring long-term sustainability and growth, as shown in Fig. (**2**).

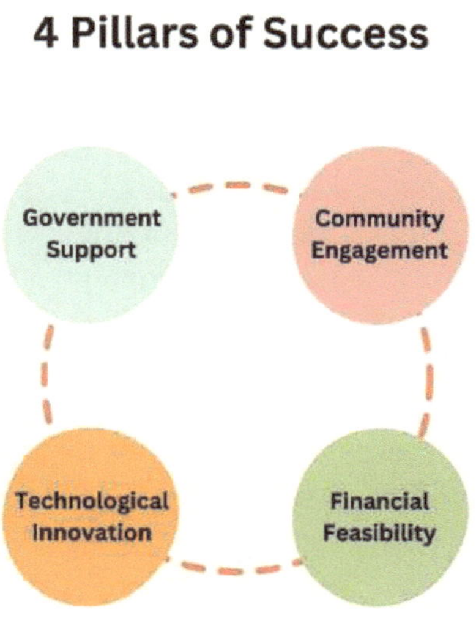


**Fig. (2).** Key factors that have been found to be responsible for the success of the sustainable projects discussed in this chapter.

## SCALABILITY RECOMMENDATION

For the scalability of the technologies discussed in this chapter, the following recommendations are suggested:

I. The government plays a key role in ensuring the success of sustainable energy projects. Their involvement will encourage the masses to actively participate and will result in more public-private partnerships.
II. Advanced modeling techniques should be leveraged to correctly predict the energy needs.
III. Community-driven approaches should be followed as they leave a positive impression on the masses, and they feel motivated regarding the authenticity of the technologies proposed.
IV. Priority should be given to tailoring local energy solutions to adjust to urban settings.

## CONCLUSION

The chapter "Case Studies and Examples" highlights the potential of sustainable energy projects in meeting global energy demands through detailed case studies. It highlights the importance of incorporating renewable energy sources in urban settings and developing community-based microgrids. In today's era, it is imperative that emphasis should be laid on utilizing renewable energy sources more than non-renewable energy sources, otherwise we will exhaust our planet's fossil fuel stocks very soon. The projects discussed in this chapter demonstrate that planning strategically, implementing innovative technology after extensive research and development, and encouraging communities for their maximum involvement can overcome challenges like regulatory barriers, financial constraints, and technical complexities in realizing sustainable development. The take-home message is that successfully implemented sustainable energy initiatives not only reduce carbon footprint and enhance energy independence by taking the burden off depleting resources to fulfill basic needs but also promote economic prosperity, social fairness, and resilience, thereby making way for a sustainable future. The readers would relate that four factors have emerged as key pillars behind the success of the projects discussed in both case studies. These are supportive government policies, community engagement, technological innovation, and financial feasibility, as shown in Fig. (**2**). These four elements are interlinked and interdependent. Government support establishes a positive opinion among the community towards new technologies. It also ascertains that the proposed initiatives do not suffer any financial barrier. Innovative technologies have characteristics of being effective and budget-friendly, which please both the government and the masses. Moreover, if the proposed project does not burn a hole in the pocket, it will always be welcomed by the government and the people. Finally, if the masses are aware and enthusiastically support the change, it has a positive impact on the governing authority, and their policies are more friendly and supportive. Thus, science, people, and the government play a significant role in achieving sustainable development.

## REFERENCES

[1] *"Framing and Context," in Global Warming of 1.5°C.* Cambridge University Press, 2022, pp. 49-92. [http://dx.doi.org/10.1017/9781009157940.003]

[2] Q. Hassan, S. Algburi, A.Z. Sameen, H.M. Salman, and M. Jaszczur, "A review of hybrid renewable energy systems: Solar and wind-powered solutions: Challenges, opportunities, and policy implications", *Results in Engineering,* vol. 20, p. 101621, 2023. [http://dx.doi.org/10.1016/j.rineng.2023.101621]

[3] M. Warneryd, M. Håkansson, and K. Karltorp, "Unpacking the complexity of community microgrids: A review of institutions' roles for development of microgrids", *Renew. Sustain. Energy Rev.,* vol. 121, p. 109690, 2020. [http://dx.doi.org/10.1016/j.rser.2019.109690]

[4] S. Gallardo-Saavedra, A. Redondo-Plaza, D. Fernández-Martínez, V. Alonso-Gómez, J.I. Morales-Aragonés, and L. Hernández-Callejo, "Integration of renewable energies in the urban environment of the city of Soria (Spain)", *World Development Sustainability,* vol. 1, p. 100016, 2022. [http://dx.doi.org/10.1016/j.wds.2022.100016]

[5] F. Fagbemi, T.T. Osinubi, and O.A. Adeosun, "Enhancing sustainable infrastructure development: A boon to poverty reduction in Nigeria", *World Development Sustainability,* vol. 1, p. 100006, 2022. [http://dx.doi.org/10.1016/j.wds.2022.100006]

[6] F. Bénard, and B. Malet-Damour, "Assessing potential of plastic waste management policies for territories sustainability: case study of Reunion Island", *World Development Sustainability,* vol. 1, p. 100030, 2022.
[http://dx.doi.org/10.1016/j.wds.2022.100030]

[7] F. Zhang, L. Zhou, Z. Wang, C. Lv, Q. Zhang, J. Wang, J. Zhang, and Y. Zhang, "Empowering urban energy transition through data-driven decision-making: A statistical examination of technological innovations in transportation and mobility", *Sustain Cities Soc.,* vol. 106, p. 105374, 2024.
[http://dx.doi.org/10.1016/j.scs.2024.105374]

[8] H. Karunathilake, P. Perera, R. Ruparathna, K. Hewage, and R. Sadiq, "Renewable energy integration into community energy systems: A case study of new urban residential development", *J. Clean. Prod.,* vol. 173, pp. 292-307, 2018.
[http://dx.doi.org/10.1016/j.jclepro.2016.10.067]

[9] T. Wilberforce, A.G. Olabi, E.T. Sayed, M. Mahmoud, A.H. Alami, and M.A. Abdelkareem, "The state of renewable energy source envelopes in urban areas", *International Journal of Thermofluids,* vol. 21, p. 100581, 2024.
[http://dx.doi.org/10.1016/j.ijft.2024.100581]

[10] R. Ge, S. Xu, M. Ullah, and P. Mark, "Assessing the impact of urban planning policies on renewable energy: A case a China using the DID estimation model", *Heliyon,* vol. 10, no. 5, p. e27099, 2024.
[http://dx.doi.org/10.1016/j.heliyon.2024.e27099] [PMID: 38463897]

[11] L. Saganeiti, L. Fiorini, T. W. Sanchez, and G. Tedeschi, *Integrating Urban Energy Resilience in Strategic Urban Planning: Sustainable Energy and Climate Action Plans and Urban Plans in Three Case Studies in Italy,* 2024.
[http://dx.doi.org/10.3390/land13040450]

[12] Adedayo Adefemi, Q.S.N. Zamathula, A.A. Umoh, K.I. Ibekwe, V.I. Ilojianya, and A. Adefemi, "Solar power integration in Urban areas: A review of design innovations and efficiency enhancements", *World Journal of Advanced Research and Reviews,* vol. 21, no. 1, pp. 1383-1394, 2024.
[http://dx.doi.org/10.30574/wjarr.2024.21.1.0168]

[13] N. A. Vukovic, and D. E. Nekhorosheva, *Renewable Energy in Smart Cities: Challenges and Opportunities by the Case Study of Russia,* 2022.
[http://dx.doi.org/10.3390/smartcities5040061]

[14] Z. Yu, and D. Gibbs, "Green industry development and urban sustainability transitions in China's latecomer cities: the case of Dezhou", *Chapters,* no. Nov, pp. 355-368, 2023.
[http://dx.doi.org/10.4337/9781803922041.00032]

[15] L. Chen, *Green building practices to integrate renewable energy in the construction sector: a review,* 2023.
[http://dx.doi.org/10.1007/s10311-023-01675-2]

[16] M. Musolino, G. Maggio, E. D'Aleo, and A. Nicita, "Three case studies to explore relevant features of emerging renewable energy communities in Italy", *Renew. Energy,* vol. 210, pp. 540-555, 2023.
[http://dx.doi.org/10.1016/j.renene.2023.04.094]

[17] P.K. Pandey, R. Kumar, V. Gupta, A. Kanungo, and S. Diwania, "An innovative hybrid controller-based combined grid-connected hybrid renewable energy system", *Electr. Eng.,* vol. 106, no. 6, pp. 7055-7072, 2024.
[http://dx.doi.org/10.1007/s00202-024-02363-2]

[18] G. Ulpiani, N. Vetters, D. Shtjefni, G. Kakoulaki, and N. Taylor, "Let's hear it from the cities: On the role of renewable energy in reaching climate neutrality in urban Europe", *Renew. Sustain. Energy Rev.,* vol. 183, p. 113444, 2023.
[http://dx.doi.org/10.1016/j.rser.2023.113444]

[19] A. T. D. Perera, S. Coccolo, and J. L. Scartezzini, *The influence of urban form on the grid integration of renewable energy technologies and distributed energy systems,* 2019. [http://dx.doi.org/10.1038/s41598-019-53653-w]

[20] H. Kaur, and R. Malhotra, "Stochastic analysis of a two-unit standby autoclave system with inspection and varying demand", *Journal of Mechanics of Continua and Mathematical Sciences,* pp. 147-163, 2024. [http://dx.doi.org/10.26782/JMCMS.SPL.11/2024.05.00011]

[21] P. Sharma, D. Bhattacharjee, H. D. Mathur, and P. Mishra, *Dual layer energy management model for optimal operation of a community based microgrid considering electric vehicle penetration,* 2024. [http://dx.doi.org/10.1038/s41598-024-68228-7]

[22] E. Khoirunnisa, *A Reference Model of Operation and Maintenance (O&M) Strategies for Community-based Photovoltaic (PV).* Microgrids on Small Islands, 2024.

[23] D. Brown, N. Dahlan, and C. Richardson-Barlow, "Evaluating microgrid business models for rural electrification: A novel framework and three cases in Southeast Asia", *Energy Sustain. Dev.,* vol. 80, p. 101443, 2024. [http://dx.doi.org/10.1016/j.esd.2024.101443]

[24] M. Eklund, K. Khalilpour, A. Voinov, and M.J. Hossain, "Community microgrids: A decision framework integrating social capital with business models for improved resiliency", *Appl. Energy,* vol. 367, p. 123458, 2024. [http://dx.doi.org/10.1016/j.apenergy.2024.123458]

[25] Y. Zahraoui, T. Korõtko, A. Rosin, T.E.K. Zidane, H. Agabus, and S. Mekhilef, "A Competitive Framework for the Participation of Multi-Microgrids in the Community Energy Trading Market: A Case Study", *IEEE Access,* vol. 12, pp. 68232-68248, 2024. [http://dx.doi.org/10.1109/ACCESS.2024.3399168]

[26] E. Etienne, and P. Robert, "Can isolated microgrids be viable? A longitudinal study of long-term sustainability in rural Senegal", *Energy Res. Soc. Sci.,* vol. 111, p. 103476, 2024. [http://dx.doi.org/10.1016/j.erss.2024.103476]

[27] M. Muttaqee, M. Furqan, and H. Boudet, "Community response to microgrid development: Case studies from the U.S", *Energy Policy,* vol. 181, p. 113690, 2023. [http://dx.doi.org/10.1016/j.enpol.2023.113690]

[28] M. Zeyad, S.M.M. Ahmed, S. Hasan, and D.M. Mahmud, "Community microgrid: an approach towards positive energy community in an urban area of Dhaka, Bangladesh", *Clean Energy,* vol. 7, no. 4, pp. 926-939, 2023. [http://dx.doi.org/10.1093/ce/zkad027]

[29] M.K. Singla, M. Safaraliev, J. Gupta, M. Aljaidi, I. Odinaev, R. Kumar, and A. Abdel Menaem, "Optimizing proton exchange membrane fuel cell parameter identification using enhanced hummingbird algorithm", *Int. J. Hydrogen Energy,* vol. 89, pp. 342-352, 2024. [http://dx.doi.org/10.1016/j.ijhydene.2024.09.211]

[30] F. Alfaverh, M. Denai, and Y. Sun, "A Dynamic Peer-to-Peer Electricity Market Model for a Community Microgrid With Price-Based Demand Response", *IEEE Trans. Smart Grid,* vol. 14, no. 5, pp. 3976-3991, 2023. [http://dx.doi.org/10.1109/TSG.2023.3246083]

[31] M. Eklund, K. Khalilpour, A. Voinov, and M.J. Hossain, "Understanding the community in community microgrids: A conceptual framework for better decision-making", *Energy Res. Soc. Sci.,* vol. 104, p. 103260, 2023. [http://dx.doi.org/10.1016/j.erss.2023.103260]

[32] C. Her, D.J. Sambor, E. Whitney, and R. Wies, "Novel wind resource assessment and demand flexibility analysis for community resilience: A remote microgrid case study", *Renew. Energy,* vol.

179, pp. 1472-1486, 2021.
[http://dx.doi.org/10.1016/j.renene.2021.07.099]

[33] G. Prevedello, and A. Werth, "The benefits of sharing in off-grid microgrids: A case study in the Philippines", *Appl. Energy,* vol. 303, p. 117605, 2021.
[http://dx.doi.org/10.1016/j.apenergy.2021.117605]

[34] A. Garg, and R.K. Ratnesh, "Solar cell trends, and the future: A review", *J. Pharm. Negat. Results,* vol. 13, pp. 2051-2060, 2022.
[http://dx.doi.org/10.47750/pnr.2022.13.S06.268]

[35] M. Kumar Yadav, R. Kumar, R. Kumar Ratnesh, J. Singh, R. Chandra, A. Kumar, V. Vishnoi, G. Singh, and A. Kumar Singh, "Revolutionizing technology with spintronics: devices and their transformative applications", *Mater. Sci. Eng. B,* vol. 303, p. 117293, 2024.
[http://dx.doi.org/10.1016/j.mseb.2024.117293]

# SUBJECT INDEX

## A



## B



## C



**Reetu Malhotra, Pooja Mahajan, Manish Singla & Ramesh Kumar (Eds.)**

## D



## E



## F



## G



## H



## I



## L



## M



## N

## P



## R



## S



## T



## W

www.ingramcontent.com/pod-product-compliance
Ingram Content Group UK Ltd.
Pitfield, Milton Keynes, MK11 3LW, UK
UKHW060011051026
15591UKWH00002B/144
* 9 7 9 8 8 9 8 8 1 4 8 2 3 *